Tales of the Dolly Llama

One Teacher's Long Journey to a Kind of Enlightenment, on which he encounters The Dowager Empress, The Better Baby Institute, the Prussian Army and more

Guy Kuttner

Outskirts Press, Inc.
Denver, Colorado

Outskirts Press
http://www.outskirtspress.com

ISBN-13: 978-1-4327-0645-6

Printed in the United States of America

When The Student is Ready, the Teacher Will Come
One Teacher's Thoughts on Education

Long ago, tucked securely away in the dim, dank recesses of boundless time, before the earth was fully formed, hurtling through a dark, frigid void in space, I was a student in college. I was back home in Chicago from my high-powered, high profile college ("The Best and the Brightest that Money Can Buy") for the summer, digging ditches on a construction crew by day, stalking the streets of the city by night.

One night my maunderings led me to a large, imposing limestone temple, a Taj Mahal of a joint, massive oak doors locked, sealing out the night and needy patrons. Huge granite blocks flanked the doors and the sweeping stone steps. I seated myself on one of these blocks, leaned back, and proceeded to light a rum-soaked crook. A rum-soaked crook is a cigar, not a Chicago politician as you might suspect. When I got to college I stopped shaving and bought a box of crooks, the better to serve my free-thinking aspirations, or appearance of liberation. The accessories make the man, you know.

So there I am, sitting on a block of stone on a warm Chicago night, meditating on infinity (or, in the vernacular of the day, 'spacing out'), when my reverie is broken by the thin, metallic clatter of little feet. Looking down I see a healthy Chicago cockroach scuttle across the stone. The roach pauses and, mindlessly, I reach my cigar over and bring it to the roach's butt. He jumps, scurries, pauses, and I buzz his butt again. This is unusual behavior for me. I was generally a kind, thoughtful kid, although I certainly had been operantly conditioned with regards to roaches. So, thoughtlessly, languidly, I burn him again. He scampers to the edge, pauses, looks back over his wing at me, and disappears over the edge of the block. Nothing strange here, a roach at night in Chicago is not really a bug to remember, but there was something about the way he had attempted to engage me that I found intriguing. So I follow.

I lean over the face of the stone block to watch. As my eyes accustom to the dim light, the roach disappears, then reappears, then disappears again. I realize that it is crawling in and out of markings on the face of the block. Angular runes, no, letters, something had been carved/written in the stone. I had been by this temple dozens of times in the past and had never noticed words in the stone. But this time I squint, follow the bouncing ball of the roach and read the words. They say, "Have we not all one Father, hath not one Spirit created us all?"

I froze in mid-drag, dropped my stogie, and stared transfixed at this primordial bug, this cockroach. Rooted, stunned, overwhelmed, I stood on that spot for a long while, trying to fathom what had just occurred, trying to accept a bug, no, a cockroach, as my teacher. I stood on that piece of pavement, on a balmy summer night in Chicago, oh so many nights, so many years ago, and took an oath. I swore to treat all creatures with dignity and respect, all creatures great and small, and to never again participate in a system that did not recognize that as its highest priority. I vowed to never participate in a system that caused distress to others, which treated others with disrespect.

Upon entering the public school system as a teacher, I have been struggling, with greater and greater difficulty, to keep that promise.

Preramble
Tales of the Dolly Llama

Hey, Guy, could you give me a hand, bud?"
I was hailed in the hall by John, a fellow new-to-the-staff colleague. I had good rapport with John. He seemed to be pleased with my treatment of Karis, his daughter, my first grade student, the light of his life. I greeted him without the apprehension I reserved for my Right-to-Phonics parents.

"What's up?"

"Well, we're looking for a Christmas present for Karis, and she seems set on some stuffie you're always talking about in class. But we can't seem to find it. Can you help us?"

Now I talk about a lot of stuff in my class, but stuffies? I was stumped. "Can you describe it to me?"

"It appears to be some kind of talking doll, an animal of some sort. I forget, an alpaca or a camel or some such thing."

"Doesn't sound familiar. Are you sure?"

Now one thing both John and I know about Karis is that she is that rare and rather frightening student who remembers absolutely everything. And accurately.

"Okay, maybe not a camel. But a doll of some sort, like a llama."

Lights went off in my dim teacher brain. Of course, a llama. Every morning as my 27 first graders walk into class, they all approach me, most in proximal, if not, physical contact, and proceed to all speak to me at the same time, much like the Tower of Babel or the New York Stock Exchange. I had developed a response, which ran something like this:

"The Dalai Llama, and perhaps a few other Realized Beings, can carry on 27 discrete and distinct conversations at the same time. I, however, not yet being fully realized, can, at best carry on two, preferably one."

Of course, a Dolly Llama. Karis wanted a Realized Stuffy she could converse with at a high level. I shared this insight with John, who took great pleasure in this revelation, as only another teacher could.

Many years have passed and I am still an Unrealized Being, capable of carrying on intelligent conversation with, at most, one person at a time. I can, however, share some "Tales of the Dolly Llama."

Chapter 1
The History of the World

Caterwaul: To utter long, wailing cries, as cats in rutting time.
Hilarity: Exuberant merriment, sometimes verging on rambunction.

I've had several decades to reflect upon my innocent declaration that I, first year teacher, planned on producing a video, written, directed, and performed by my 5th/6th grade class, entitled, simply enough, "The History of the World: a student perspective."

Okay, so perhaps my choice of the faculty lounge as the venue for that innocent declaration may have been a mite inappropriate, a mite outré, a bit over the top. But I was totally ill-prepared for the caterwaul of hilarity that greeted my admission. In considerable retrospect, I probably should have courted a private ignominy, played my certain failure closer to my vest. Maintained a poker-faced front as I reviewed my preordained dead man's hand, self-dealt. Kept my hubristic stupidity to myself.

But that was not, and still is not, my way. So, smiling

and upbeat, I strode into the teacher's lounge and announced my certain doom. Accustomed to a gentler, headshaking disbelief from my "peers", I was rattled by the uncontrolled response: belly heaving, diaphragm gripping, collective gasping for breath. Drinks spilled, paper cutter poised in mid descent, mimeo paused, disbelief hung suspended by a gauzy thread. Attempting an outward calm I did not feel, I backed slowly to the door, groping uncertainly behind me for the doorknob, face frozen in a rictus of good humor. Forgetting that the door opened in, I was whacked in the back of the head by the heavy door as it was flung open by Rich, the next teacher seeking faculty refuge.

Sprung from the trap, I gracelessly made my escape, hearing, as the door closed behind me, "You won't believe what Kuttner's doing now..."

In abject retreat, I regrouped. "Well, perhaps the topic's a bit broad," I admitted to myself, "but what about the History of the United States? We ought to be able to pull that off." Back to the drawing board I limped, not willing to cast my lot with the old burnouts, those unwilling to take a risk. "What's the worst that could happen?" I asked myself.

I was to find out.

In the weeks that followed, batteries ran down, cables were misplaced, entire scenes were "recorded" with the camera on 'pause', old technologies failed, new technologies were scrapped. And those were only the technical difficulties. They paled alongside the human difficulties. Oh, teacher, thy name is foible, Foible Kuttner.

In my ethical drive to make no decision that would hamper creativity, I gave few, if any, directions. None that were comprehensible to ten and eleven year olds, anyway. Props? I'll get the materials. Dialogue? Use your intuition. Historical accuracy? Be creative, incorporate multiple perspectives, a cubist view of reality.

It was, I think, the third week of filming when the wave of fatigue swept me off my feet (never turn your back on the ocean, or a sea of kids). As little Jerry decided to debark Washington's boat and leap from ice floe to ice floe, desk top to desk top, across the frozen Potomac, I felt my legs go out from under me. Shuffling, stumbling back into the corner of my classroom, I slid down the wall, burying my head in my hands, cradled between my bent knees. I had literally and figuratively hit the wall.

How long I huddled there, I don't know. Certainly long enough for Washington's ragtag army to defeat the British at Trenton. Long enough for Cornwallis to surrender at Yorktown. I might be there still, had Christa not noticed me, slumped, defeated, and hyperventilating in the corner. With a gentle hand on my shoulder, she attempted to rouse me from my morass. "Are you okay?" she whispered in my ear.

"Is he dead?" little Jerry asked, not totally devoid of hope.

"No, I think he's tired," Jesse ventured.

"Or depressed," suggested Qué.

"No, I'm alright," I mumbled, staggering to my feet. Like a punch drunk boxer convincing my trainer I could go another round.

Slowly, reluctantly opening my eyes, I saw thirty school kids eyeing me warily, some concerned, some amused. Surveying the bales of tea still floating about the class from last week's Boston Tea Party, I managed a strained smile. "Now, where were we?"

****

By the way, the Americans didn't win the Revolutionary War. It was the kids. The kids won. And the rest, as they say, is History.

3

Chapter 2
The Pickle Doesn't Fall Far From the Vine
A Brief History of Education

N ow this is a fine pickle you've gotten us into, Stanley."
Whenever Laurel and Hardy, old vaudeville and early
film comedians, found themselves waist deep in the big
muddy, or some other substance, Oliver Hardy would turn
to Stan Laurel and blame him for their predicament.
Education, especially public, is indeed in a fine pickle, but
there is no easy scapegoat. No evil dark overlord (Dr.
Stupid) moved all of us witless game pieces into this losing
configuration on the board of life. No one particularly likes
the pickle we're in (with the possible exception of some
sadistic administrators), but how did we get here?

I was so impressed with the sales popularity of Stephen
Hawkings' "A Brief History of Time", that I thought the
public ready for a follow up companion piece on the literate
coffee table, "A Brief History of Education."

I figure, reasonably enough, that unless we develop a

sense of where we came from, and where we are, we can't possibly have the vaguest clue as to where we're going. So, how did we get here?

7 million years ago, when humans diverged from apes, in a modest place no more nor less auspicious than any other spot on the planet, the Olduvai Gorge in Tanzania (or thereabouts), there was no fanfare, no trumpets, no presidential proclamation. We observed our cousin apes, saw what was successful for them, and copied it (even before Xerox machines!). Their schoolhouse was the Great Plains of Africa, their class size 1 or 2 (the original class size reduction), their Best Teaching Practices were modeling and Hands-on Learning. Basically, I'll keep doing what I always do, because it makes sense. You watch me do what makes sense then, when you're ready, you help me, or do it on your own. Feedback is immediate. If the termites all bite you on the ass instead of getting lodged in your Termite Extraction Device, you've probably messed up. I call this Avuncular Education, sensible education by aunts and uncles who care about you, and who are doing obviously important stuff.

So, was this Avuncular Education successful? You've got to be kidding. After mastering our craft of humanness, we erupt upon the earth in a flood of fecundity, a tsunami of technology, (a wave of terror?), unlike any ever seen. Yes, the basic form of education: take a stick, poke around, see what happens, if you like it, do it again, if you don't, don't, do something different, is very effective. It's the Scientific Method. It's not something you are taught, it's something you Discover.

So, it works. It is dizzyingly successful for 6.99 million years. So what's the problem? The problem is called the

Industrial Revolution. Our previous educational system had been geared to learning and doing things that made sense, things that dealt with personal and social survival. If our actions harmed ourselves, the social fabric, or the environment, there were harsh negative consequences. But with the rise of Conquest, Mercantilism, and Industry, we no longer focused on doing stuff that made sense, we began to focus on doing stuff that would help a small number of people accumulate stuff. My wife calls it, Stuffism. So how did Stuffism shape the educational system we diffidently call our own, today?

With the rise of Stuffism, came a demand in the Western world, for efficiency. Folks in authority thought that kids were not learning much during their time in school; they might as well be on the assembly line, producing more stuff. Authorities thought there was a lack of organization in the schools, the kind of organization that allowed the Industrial Revolution to take off. And of course, with organization comes Management, the ethical and financial imperative to manage other people, to boss them around. There was a search for a model of education, a model of what schools, teachers, and kids should look like. One was found: the fiercest, most efficient fighting machine in Europe: the Prussian Army.

In Chapter 4 we'll investigate how the Prussian Army of the early Nineteenth Century matured into the public school system of today.

Chapter 3
Take Me Out To the Ball Game

By independent audit, classroom teachers make from 1,200 to 1,500 decisions each and every day. This resolves to about 3 to 4 decisions a minute. In my second year of teaching, and in every successive year, I was to definitively demonstrate that those decisions were not all correct, nor were they all wise, nor even smart.

The setting was the ball diamond at school. My third and fourth grade kids were arrayed for a seemingly innocent game of kickball. My intent was to insure that all kids played equal amounts, with equal opportunity for success to all. In my mind, this meant that every kid batted (kicked) every inning, outs and score held as inconsequential.

As the ultimate arbiter of fairness, I pitched, which consisted of rolling a large, schoolyard bouncy ball toward the plate. Kids who had not yet mastered the skill of balancing on one foot and kicking with the other were accommodated by slower pitches as I inched my way slowly towards the plate. Melissa was up, outfitted in awkwardly fitted gel shoes, sandals really. As she clutched her skirt with her hands, she spasmodically flailed away at the approaching ball, unable to accurately time the approach. This meant, in reality, that she began to reflexively kick even before I had released the ball, repeating this clumsy gesture as if she were an insect trapped on flypaper, with only one limb free.

My ceaseless encouragements, "Attaway, girl, keep your eye on the ball, take your time, patience, patience," were punctuated with attempts to keep the fielders from inching forward so far they could almost reach out and touch her with their hands. "Back up, give her room, let her kick, give her a chance. Joey, back. Richy, back. Joey, further." Like jackals sensing hapless prey, they moved in for the kill.

Jessica, standing impatiently in line, pushing and shoving for a clear view of the slaughter, issued her judgment. "It's not fair!"

"What did you say?" I challenged, drawing myself up to my full teacher/coach/umpire stature. Now, as everyone, even teachers know, "What did you say?" is not an honest question meant to ascertain what it was that someone had actually said. It is a two-pronged offensive to: one, buy time, and two, give the opponent an opportunity to recant. My ploy failed.

"It's not fair! And it's stupid!" Jessica responded, not giving an inch.

"Jessica, get out of the line."

"Fuck you!"

Wow! Needless to say, this was an escalation of the arms race I hadn't anticipated. A dangerous destabilization of the balance of power.

Fifty little eyes were mutely riveted on me, sharing my sense of flabbergast, fascinated how I would deal with this unthinkably powerful return of my weak serve. Oh, how I wanted to ask the kids what I should do, what would be an acceptable response, how to gracefully step back from the brink of personal and professional annihilation.

At such times, primal instinct takes over, and, for most males, unhappily, the flight or fight mechanism shifts into overdrive. I exhaled (I must have been holding my breath), placed the ball on the mound, and strode purposefully toward Jessica. She (bless her little combative heart) stood her ground. I reached her, bent over, swung her on my shoulder, and spun around, striding across the field. It wasn't until I reached second base that I realized what I had done, and where I was going: To the office to pass the buck.

Jessica, meanwhile, kicked her feet and howled threats at me, "Put me down! I know my rights! I'll sue!" Personally, I didn't give a damn. I knew neither her rights nor mine, for that matter. I only knew that I had an ornery foul-mouthed, fourth grade girl on my shoulder, playing out a scene which neither she nor I had scripted, providing unforgettable entertainment and horror, both, to twenty-five eight and nine year old kids. And I was stuck. The time for a graceful retreat for either of us was long past. We simply had to play our preordained roles.

Over the years I've wondered, what if there had been no office to haul her to, what if there had been no safe and easy depository for her? Perhaps I would have carried her home. Perhaps I'd be carrying her still, kicking and screaming, loudly proclaiming her rights.

I was Atlas, carrying the weight of the world on my shoulders. I was the world, hauled helplessly about. I was Sisyphus, condemned to roll the boulder up the mountain. I was the boulder.

And I still am. Actor and acted upon. Such is the role of teacher.

Chapter 4
The Prussian Connection
Or what's good for the military is good for the nation

When we left off, we had just fast forwarded through the first roughly 7 million years of human (or proto human) history. This leaves us at the year 1805, or 99.715% into the course of humanness on the planet. Why pick up at 1805? Well, that's the year that Napoleon and his armies crushed the powerful Prussian forces. Following the humiliating peace treaty in 1807, the defeated Prussians vowed to never again face such loss of face (hell hath no fury like a Prussian scorned).

Prior to this time, most European armies consisted of the social dregs. Consigned to a degrading life, cast aside by the 'progress' of the Industrial Revolution, they were a rag-tag, rebellious clutter of humanity, conscripted into service (military slavery) only during times of war. They were poorly led, fed, housed, clothed and ill-trained for the task at hand. Not surprisingly, they didn't do very well.

The Prussians, however, fresh from humiliation by the French, remade their armies from scratch. They adopted modern-day factory methods to create the Modern Soldier: professional, meticulously trained, and frighteningly successful. In 1813 and again in 1815, they defeated the Napoleonic forces and, in stunningly ruthless and efficient fashion, were successful in wars against Denmark, Austria and France in the 1860's and 1870's. The Prussian military was the envy of the continent. How had they achieved this turnaround?

The Prussian military studied the factory models of the Industrial Revolution, as well as the farm models of the Agrarian Revolution. The factories had subjugated the workforce such that free will, free thought, and innovation were squelched. They substituted a workforce that would uncomplainingly perform the same small task, over and over, to produce whatever stuff was required. Input raw material, manipulate it in exactly the same way, and output a standardized product, free of 'flaws' or deviations. The farms had begun to pen animals, developed the first feed lots, and imprisoned hens in the first barracks, controlling the messy animal component of husbandry. Efficient, standardized and predictable; a miscegenetic mating of Motel 6 & Kentucky Fried Chicken.

To achieve this goal of standardization, the Prussian military selected their recruits, grouped them by age, height, weight and experience, and housed them together. They were subjected to intense, personality- destroying training (creating the *tabla rasa* or clean slate), endlessly drilled and tested. Weaklings were weeded out. The result? A completely cohesive, standardized unit that moved in unison at the whims and directions of one man: the commander. An awesome fighting machine, unhampered by divergent thought or ethical hesitation.

This is the model that stormed the Prussian school house, forever changing the face of education. Gone was the simian, proto human model of avuncular education (monkey see, monkey do). Never again would a bumbling teacher oversee a rather disorganized array of kids of all ages, learning (or not) at their own paces, the older kids tutoring the younger. Now the kids would be classified by age, gender, and ability. They would be subjected to intense, personality-destroying training, endlessly drilled and tested. Weaklings would be weeded out. Students (recruits) would be grouped and marched, lockstep, grade by grade, toward graduation. The product (the kids) would be standardized to suit the needs of the culture: an awesome productive force, prepared to produce lots of stuff, over and over again, unhampered by divergent thought or ethical hesitation. A modern efficient workforce.

With typical Prussian efficiency, this new model of education spread throughout Prussia, then the Continent, within a decade or two. American 'reformers' caught scent of the winds of change, eager to import these modern methods to a young, vaguely anarchic and notoriously inefficient United States. Horace Mann, founder of the Massachusetts Board of Education, the first state board in the land, was a particularly eager proponent of the Prussian model (an unselfconfident America tended to think that anything European was better than anything homegrown). And with youthful American eagerness and lack of forethought, the Prussian-American Connection was forged.

Chapter 5
Turkey Butts

Click, click, squish, squish, the telltale treads of Amber's stiletto heels and Sarge's sensible nursing shoes echo down the hall. Sarge, in her floral uniform appears at my open doorway first, followed by Amber in her tight, slit sheath. Sarge: "Thanksgiving's coming up. I'm the Pilgrims, Amber's the Indians. Do you want to be included in the feast?"

Jeez! I glance helplessly over at the Seasonal File Cabinet I had inherited, organized by Holidays. I hadn't even sprung into Fall, let alone addressed Halloween with the vigor expected of a First Grade teacher.

"Sure," I mumble, not at all sure. "What's left?"

"Well," Sarge replies, as uncertain of me as I am of myself, "You could split your class up, half Pilgrims, half Indians, and join us."

Having vaguely anticipated I would spend Thanksgiving exposing the rapacious colonialism of Columbus to my six-year olds, I offer halfheartedly, "Why don't we make the turkeys?"

Sarge and Amber exchange one of those knowing looks that, although warm and avancular, makes me feel as if I had worn my underwear outside my pants. "Okay," they chirp, "That'll be different."

Click, click, squish, squish, I hear them retreat down the hall, having done their good deed of including the village idiot in the festivities. They're trying to make up for the Spelling Bee, I tell myself. Two weeks before, I had received in my mailbox a list of the Official Words for the school spelling bee. "What do I do with this?" I had asked my colleagues.

"Oh, we don't worry about that in First Grade," they had assured me.

Good, I had thought. My spelling program had so far consisted of informing my charges that, "Some of the most intelligent people on the planet can't spell their way out of a paper b-a-g. Just write down what it sounds like to you."

One week later my kids had been skunked, subjected to a very public humiliation in the school gym, while Sarge's and Amber's kids had all scored 100%. Set up, I thought, I had been set up.

Okay, so I would accept their peace offering. Turkey it would be. Now, I thought, how the hell do you make turkeys?

As was my wont, in those early years, I proceeded without a model, so I wouldn't stifle the kids' creativity. We blew up balloons, we ripped newsprint into strips, we fabricated our own paper maché, we made individual turkey drying stands out of two by fours and PVC pipe, we set to work on feathers with construction paper and scissors, we dove into the turkey production mode amidst a sea of paste, songs, and general frivolity.

Time passed. Thanksgiving drew near. When I realized the turkey bodies would not dry (perhaps I should have used a recipe) desperation turned to despair. And despair, of course, is the mother of teaching ingenuity. Knowing there was no longer time to make and affix necks, wattles, beaks, and heads, I lifted all the bodies off their stands and sawed them in half. When the kids came in the next morning, the day before the feast, I excitedly announced that we'd be making turkey hats. We would play the role of turkeys, ourselves.

Never having been turkeys, consciously at least, the kids became as excited as I. We hurriedly took our pre-made feathers and glued them all over the bodies, creating a fan of feathers for the rear. A bit of elastic and we had our costumes, just in time to welcome our guests. We dragged all the desks out into the hallway, laid long strips of bulletin board paper on the floor, and almost comfortably seated ninety first graders on the linoleum, stretching wall to wall.

Amber, Sarge, and I leaned against the door with pleasant weariness, observing the cooperative feast. Pilgrims, Indians, and Turkeys joyfully passing paper plates of potatoes, stuffing, and weinies, Pilgrims and Indians chattering, turkeys gaily gobbling away (my class had made a pact not to speak, only gobble). I was explaining the turkey debacle to my colleagues, noting there had not been time to make heads, necks or wings. "So," ventured Amber, "So those are..."

"That's right," I said, "Those are turkey butts."

As Sarge and Amber exchanged another of those now familiar looks, my gaze passed fondly over my kids, gobbling gleefully. Turkey butts or not, they were having a grand time.

And, for not the last time, I acknowledged, "God, I love First Grade."

Chapter 6
America Goes Prussian
Or The Prussians are Coming,
The Prussians are Coming!

Nineteenth Century Europe: The Prussian army is the baddest fighting machine on the continent. The Prussian schools are busy pumping out cadets, er, students, as identical as they can make them. These perfect products are themselves producing in the military, in the factory, and on the newly revamped farms. The Industrial Revolution is a success.

Across the pond, Americans are eyeing the Prussian success with envy. Horace Mann leads the charge to import Prussian educational reforms to the U.S. To give old Horace credit, however, he is attracted not only to the efficiency and standardized output of the Prussian schools, but also to a kinder, gentler component. This aspect of Prussian schooling was represented by the Swiss philosopher/educator Johann Pestalozzi. His view of learning/teaching was child-centered, with the emphasis on individualized instruction and development, fostered in a nurturing, familial environment.

Unfortunately, a powerful cadre of Bostonian educators opposed this 'soft education', arguing that schooling should be hard, in order to build character. Pleasure had no place in a system whose purpose was to produce tough, efficient workers. This view won out. Structure and standardization prevailed, and the hard-edged Prussian school system was poised to conquer the New World.

Prussian organization was wed, in a joyless matrimonial bond, to the prudishness and conventionality of the Victorian Age. This era championed the position that learning was like sex: unpleasant, difficult, yet necessary. This view controlled American education for the rest of the century and, to a large extent, still holds us in thrall today.

Enter the Progressives. In January, 1896, philosopher John Dewey gathered 16 children together in a small, brick house on the South Side of Chicago. The Laboratory School of the University of Chicago was an experimental program to test Dewey's ideas on education, to wit: that children learn best by doing, that students should decide the curriculum, and that education is, at heart, an exercise in democracy (imagine that!). Although Dewey's views changed quite a bit over the course of his long life, still his experiment struck a receptive chord in the hearts of advocates of social change and justice, such as Jane Addams of Hull House.

Schools of Progressive Education began to spring up around the country to promote joyful environments where children's natural curiosities were engaged. Some were free, wild places where kids ruled; some where the natural curiosities interacted with an organized curriculum. Progressives changed their views and directions (John Dewey despaired over schools where kids spent the day making nut bread but could not read). But all were places that strove to break free of the 'dull recitations' of traditional schools.

At the same time, the ugly spectre of Intelligence Testing began to cast a dull, bleak shroud over the educational landscape. Intelligence testing was born of a desire to determine who among us was really fit to be educated, and to separate the intellectually elite from those who were merely trainable. Imagine what would have happened had Homo Habilis or Homo Erectus selected a few primal humans who displayed early test-taking skills, and focused the transmission of culture to those early Best and Brightest. However, at the turn of the century, this brand of Social Darwinism, the Survival of the Brightest, joined the educational debate.

And, once again, it was a war, this time the First World War (known as the Great War, or the War that would end all Wars) that catapulted this new educational reform, the repression of testing, to the forefront of American education. As Frank Smith says, "It took a war to make testing a growth industry."

We'll deal with the issue of testing in later chapters, but we've got to acknowledge that it was the success of wartime testing that fitted us all with a noose that has tightened relentlessly over the last century.

Chapter 7
Respect

Standing at my open classroom door on the first day of school, I welcomed my new students to the room. It was my first year in fourth grade, having been elevated from first grade teacher status. The previous spring, the Superintendent had called me into her office and offered a change in grade level. "You've done a good job in first grade," she smiled at me, "But there are openings next year in first, fourth, or seventh. Do you want to make a shift?"

"Fourth," I blurted.

"Take your time. There's no rush. You may want to think it over."

"Fourth," I repeated. Junior High intimidated me, and although I had enjoyed first grade, the parental pressure to learn to read was immense. At first grade Open House I had revealed my ignorance of the reading process by declaring to parents, "The acquisition of reading skills is little understood by the educational community. I consider it rather magical. Kids can be exposed and their natural tendencies nurtured, but the acquisition itself is shrouded in the mists of the individual mind. It's magic."

Some parents had been less than impressed with my rather ethereal assurance. As I fielded the inevitable questions from the parent community, day after day, "Is she reading, yet?" I would share with colleagues that I'd really rather teach first grade in an orphanage. That option unavailable to me, I chose fourth grade, hoping the more experienced teachers had already taught the kids to read.

"This is Jeremy. He's a handful. You can have him," says an overwhelmed parent, happily handing over her charge, and spinning on her heel without waiting for a response. "Hi, I'm Monica. This is Melissa. She's never been to school. Home schooled for four years because of the bad influences, you know. " I exude what I hope appears to be good influence and smile. "Brandon can't read. Not a lick. His teachers haven't seemed to notice." "We'll do what we can," I offer, hopefully. "I'm Phonetia, Ariana's mother. She's reading Tolstoy, aren't you dear?" Why, I think, why would a fourth grader read Tolstoy? "This is Sonja. She doesn't know any math." I beam out math confidence and thank her for the challenge.

I take note of a Native American woman and her daughter, hanging back from the throng, sizing me up. She strides up to me when the crowd disperses, most parents having retreated, several peering anxiously into the class as their children find their name cards, take their seats, and begin their first lesson of the new year, copying my neatly penned cursive message off the blackboard, "Welcome to fourth grade. Don't worry! Do the best you can, then be happy!"

"Hello," she says, meeting my eyes. "This is my daughter, Tori. She is half Yurok, half Hoopa. She is a wonderful child. We are proud of our heritage. Treat her with the respect she deserves and she will do anything for you."

"Uh, thank you. I'll do the best I can," I spoke to her

retreating back. I looked down at Tory, who gazed back at me with large, intelligent chestnut-brown eyes, and felt as if this stranger had just bestowed upon me a rare and important gift. "Welcome, Tori. Find your seat, please, and make yourself at home."

****

Years later I took my class on a camping fieldtrip to a reconstructed Indian village, built with the cooperative effort of the local Native tribes, constructed in the ancient tradition, using only authentic tools of stone, bone, and wood. Tori, having founded the Native American Club at our local high school, met us as our guide at the dance pit.

She was stunning, tall and confident, wearing her full Brush Dance regalia, anklets and necklaces of deer and abalone, magnificent dress of elk hide with hundreds of beads and abalone shell that rang as she moved. Red woodpecker scalps and bits of fur trimmed her outfit, making her look like a Nature Goddess. She spoke clearly and eloquently to the kids, delving in fine detail into the fabrication and import of each piece of her dress, constantly stressing the care and love that went into the making of each item.

"But," she cautioned the kids, "as beautiful as each of these pieces is, it is the love of Nature, and the respect for our place in the natural world, that make all these things special. You must especially show respect toward your Elders in all that you say and do. You are lucky to have a teacher like Mr. Kuttner to help teach you that respect. I am in training as a Medicine Woman and as a teacher so that I can continue that tradition."

As I looked at this powerful young woman, a tingle ran up my spine. I remembered her mother's charge to me, "Treat her with the respect she deserves," and hoped I had.

Chapter 8
The War to End All Wars, Doesn't
But it Does Cripple Progressive Education

Oh, man! Not more history! I mean, like, it's so, you know, like, old! It's like so then. It's hella then. The antipodal imperative to nowness. History is like his story, you know? I'm into mystory. What about my story?

Dear and gentle reader, just suck it up, okay? Learning is supposed to be tough.

It's the early 20th Century, Progressive Education is battling Victorian Education (learning, like sex, should be unpleasant) as well as being assailed by Vocational Education. America decides to enter the First World War, the Great War, to save those insane Europeans from themselves. For four years, the military bosses had raised their swords, and wave after wave of young men had climbed out of their trenches and into their graves. On many occasions France lost up to 70,000 soldiers a day. We're talking major insanity, here.

The year: 1917, the President: Woodrow Wilson, the campaign promise: keep the U.S out of war. So we enter the war. We draft hundreds of thousands of men. What do we do with them? How do we split them into officers and others? How do we assign jobs? Well, we employ teams of statisticians to develop tests to sort and group (slice and dice, mince and blintz) the draftees into appropriate categories.

A mass of humanity enters the military machine and a battery of tests spits out gunners, medics, clerks, and bosses. 1,700,000 recruits are tested and assessed during the war. We sort and group so successfully (remember the Prussians?) that the war ends in a year. Peace breaks out and thousands of professional testers are thrown into the civilian job market, bringing their newfound humanity-sorting skills with them.

As after previous wars, society (in general) and education (in particular) are quick to apply the lessons of the trenches. Education, faced with multitudes of 'draftees' (pupils), realizes that simply grouping and categorizing their inductees has not been successful. The results had been less than spectacular. We decide we need a tighter grip on the process and more management, greater hierarchy. MORE CONTROL! And thus dies the experiment of Progressive Education.

Yes, there remained islands of folks dedicated to humane, more child-centered education. A Colonel Parker took over the Quincy, Mass. school system and threw out the division of subjects, grades, rules, and prizes for being good. The entire city of Gary, Indiana embarked on a noble, progressive experiment. Schools in New York City and Winnetka, Illinois carried the banner. The Progressive movement itself was torn apart by arguments that pitted the 'practical' wing (courses for living: shop, mechanics, Home Ec.) against the proponents of a liberal education (humanities, the arts, history). But the major battle was with intelligence and educational testing.

As we gear up for the invasion of Europe before our entry into World War II (we no longer refer to wars as the War to End All Wars) we rely increasingly on greater organization, hierarchy, and logistics. This model is so successful that, at the end of the war, the U.S. is left, virtually alone, as the dominant nation in the world. The power structure that allowed us to overpower our military opponents is, once again, imposed upon the educational system. Greater organization, greater management breeds an entrenched bureaucracy of educrats far removed from the classroom, far removed from teaching itself. Administrators who break out in a cold sweat if required to actually run a classroom for a few minutes. Administrators more familiar with spreadsheets of test results than they are with the kids.

Much as safe, remote, and removed Generals had sent multitudes to their deaths, a new generation of educational commanders is poised to lead generations of students toward a predictable, standardized result. The attempt to reduce civilian casualties is half-hearted, at best.

Education becomes a huge growth industry, modeled on the military and industry, with a dizzying array of administrators and middle managers. Antipersonnel weapons are traded for industry-produced textbooks and scores and scores of tests. Tests, tests, and more tests.

As Gerald Grant says, "The short history of American education in the 20th century is that (testing) won and Dewey lost."

Chapter 9
The Sounds of Silence

It reminded me of the old shaggy dog joke. Upon returning home from work, George is greeted excitedly by Martha, his wife: "George, George, you won't believe this, Rover can speak." "Well, what did he say?" "He said, 'This food isn't fit for a dog.'" Confronting Rover, George asks him, "Is this true, Rover? How come you never said anything, before?" Rover eyes his owner and replies, "I never had anything to say before."

But there was nothing funny about this situation, nothing funny at all. I had just been informed that one of my students for the next year, Maria, was suffering from a disorder called Selective Mutism. She didn't speak, not a word, not a sound, nada. Her family was Mexican, she had been born there, and had never uttered a word during her previous three years in school, neither in English nor Spanish.

In the week before school began, I read up as much as I could on Mutism, although I couldn't find a thing on Second Language Mutes. For all I knew she was a polyglot mute, mute in many, perhaps hundreds of languages! It all seemed fascinating and a bit overwhelming.

The keying mechanism appeared to be anxiety. So my job was to make Maria as comfortable as possible, this without any feedback at all. Seemed like a bit of a challenge, but that's what this job is about. Okay, we'll give it a shot.

On the first day of the year, I had no difficulty picking Maria out. I had four Mexican kids, three Native Americans (two with Mexican blood), two Azorean, and one Filipino, but Maria stood out. Or rather, she didn't. She receded. Although she stood her physical ground, she had an aura of psychic withdrawal, of not quite occupying the physical realm.

Maria was stout, sturdy, Mayan. She had blunt features, jet hair cropped short and straight across her brow, flat nose, flat face, her entire aspect was flat, with the rather startling exception of her eyes. They were deep, rich, chocolaty, the eyes of a doe, not just in appearance, but in wariness, ready to bolt, to bound off at a moment.

In the days and weeks, even months, that followed, I made little progress. Maria was evidently at ease with me, relatively speaking, even taking to shadowing me. I would come out of the staff lounge, the office, the men's bathroom, to find Maria waiting for me, sometimes seated cross-legged on the ground, staring at the door like a spaniel. It began to make me ill at ease. The mantle of Leader of the Cult, the Masked Crusader, weighed uneasily upon my shoulders. I had to make this Strange Encounter be about Maria, not about me. I had to maintain this atmosphere of cordiality, and yet somehow distance myself, so the focus could be fully on her.

I enlisted a team consisting of Darla, the Speech and Language Specialist, Renata, the Spanish/Second Language

Specialist, and myself. Darla found an Hispanic Mute in Texas and, with Renata's help, got some important strategies. We found an expert on Selective Mutism in the next county, developed strategies with her. The three of us became not so much obsessed, as highly occupied with the situation. We all felt that time was of the essence, that a breakthrough was essential to this little girl's psychic and educational health and well-being.

All the while, Maria watched us as a prey animal might track a predator, going about her business, grazing and browsing, but always, always aware of the presence of the predators. I could feel the heat of those doe eyes on my back. Turning, I would catch the slightest, most subtle movement from her, a slight downcast of the eyes, a microscopic tightening of the lip, the lazy swing of a wisp of hair, all signs that she had shifted as I turned.

This rather bizarre pas-de-deux would doubtless have continued had we not encountered a stroke of luck. One day, while Renata was calling Maria's parents, Maria herself picked up the phone. "Holà." "Holà," Renata countered, and before Maria was aware of who she was talking to, they had a conversation. Maria was stuck. She could no longer pretend she couldn't speak, she had to move forward.

We pressed the advantage. Renata called almost daily, went to her house to play with Barbies and watch movies, began calling her at school and talking to her on the phone. Still no direct speech, but indirect, at least.

How to translate this progress into the classroom? Maria was not going to make this easy for us. Believing that the eyes are the portals to the inner being, I made more and more of an effort to make eye contact. I had to be a bit sneaky, turning a bit faster than normal, positioning myself at times rather awkwardly in order to facilitate a face to face.

But the year was almost over. We had to make our move. Renata asked Maria if she would be willing to record a tape of her reading, and allow me to listen to it. Maria nodded, yes.

That Thursday, after the kids had left, I cued up the tape and listened, misty eyed and enthralled, as Maria read (in almost unaccented English!) the story of Tiffany and Barbie on a dude ranch, "Horse Trouble". Renata came into the room as the girls were dismounting after a harrowing ride, legs shaky but proud.

Renata and I looked at each other, too overcome to speak, but feeling much the same way, shaky but proud.

Chapter 10
What's Wrong With Tests?
Or, "The Pig Doesn't Get Heavier by Weighing it More Often"

So what's wrong with tests? Don't you want to know how the kids are doing? Absolutely. I want to be as up-to-date as possible on how the kids are doing. Standardized tests, however, will not tell me.

Don't you want to know if they're learning? A good teacher knows if the kids are learning. Actually, any sensitive person can tell if the kids are engaged. It's not always easy to tell *what* they're learning.

But won't tests predict how they will do out in the world? No. There has not been a single reliable study indicating that.

But won't tests predict how they will do in college? No. In fact, the UC system is phasing out the almighty SAT's, the battery of tests that rank slightly below Judgment Day, because they don't work. They don't, and never have done, do what they claimed. A billion dollar scam.

What standardized tests do, and very well, is to predict how kids will do on future standardized tests. Kids who do well on First Grade standardized tests do well on Eleventh Grade standardized tests (unless they're burned out on testing!) What tests also do is drive American education.

"Well," (asks the shill planted in the audience), "how did this come to be?"

I'm glad you asked. Standardized testing derives from intelligence testing, which derives from roots so elitist, so classist, and so racist that all but the most case-hardened skinhead has got to find it repulsive.

Henri Binet, whose name is still associated with the most famous IQ test, the Stanford-Binet, launched the 'science' of intelligence testing to determine who should gain entry to the overcrowded Paris lunatic asylums of the late 19th century. It occurred to wealthy white folks that testing could also be used to determine not only who was nuts, but who was exceptional, who was worthy of breeding. That's right, it spawned the 'science' of eugenics, the cult of the 'well-born'. Intelligence testing and Eugenics were sired in the same bed, begot of the same desire to proliferate the right sorts of people and extinguish the wrong sorts. As the Star-Belly Sneetches observed in Dr. Seuss' landmark study of prejudice, The Sneetches, "'We're the best kind of Sneetch on the beaches.' With their snoots in the air, they would sniff and they'd snort, 'We'll have nothing to do with the Plain-Belly Sort!'"

Francis Galton, cousin of Darwin and member of a wealthy, intelligent family, developed testing as a tool to study his family's 'genius'. He wanted to map the transmission of 'good' traits and to encourage 'successful' humans to reproduce, and discourage (or sterilize) 'unsuccessful' humans. This led to Social Darwinism, the belief that smart, rich white folks deserve their status (have earned it, evolutionarily) and everyone else (the dregs of humanity) deserve theirs.

Eugenics Societies sprung up all over England and the U.S. The first sterilization law was passed in Indiana in 1907. In 1985, 22 states still had these laws on their books. As proposed by the American Eugenics Society, sterilization was necessary to "eliminate undesirable types of people...including borderline persons of low intelligence and unstable temperament who are of little direct value to society." The Society referred to eugenics as 'race hygiene' (clean up that dirty culture).

Lest self-righteous Americans think the Nazis invented ethnic cleansing, be advised that tens of thousands of Americans were sterilized involuntarily from 1927 on, in response to a Supreme Court decision. Canada followed suit, sterilizing many (especially Indians) based largely on IQ test results.

But beyond sterilization, Eugenics reared a slightly more subtle head when applied to schooling. In the aftermath of World War I, there was a mass movement toward applying the modern science of testing (used so successfully to sort draftees into military career tracks) to education. Lewis Terman, the Stanford IQ King, felt that intelligence tests would prove of great use to school administrators in assigning students to vocational tracks. He claimed that, "The dull remain dull, the average remain average, and the superior remain superior." Why waste a good education on the dull? Why cast pearls before swine?

Why indeed.

Chapter II
Pagan Rights

"See me," TS.

The tersely worded note was calculated to strike nameless dread in the heart of the recipient. And oh, how successful it was. Whenever our boss, the Dowager Empress, wanted to rattle us she would pen that simple line on an orchid colored post-it note and tack it to our mailboxes. It never failed to drive fear into the staunchest heart.

The recipient would almost always collar innocent passersby, asking, "Do you know what this is about? Why would she want to see me?"

Colleagues would give us a wide berth, as if we were leprous, not wanting to be associated with the condemned, fearful whatever we had might be contagious. Much folks avert their eyes from a traffic accident thinking, on some gut survival level, "Boy, I'm glad it's not me," as though there were some limit to the amount of distress distributed in the world, as if another's distress somehow lessened your odds of the same being visited upon you.

But there was never any hope of avoiding the inevitable. I had tried, once. My mailbox constantly in a state of hopeless disarray, I reasoned that the Chief might think I had overlooked her command.

Wrong! On that occasion she had issued her ultimate demand which came in the form of yet another note, "**See me, now!**" That was bad.

I strategically planned my interrogation towards the end of my break, so I'd have an excuse to beat an early retreat. "Hi, chief," I said jauntily, "What's up?"

The boss looked up from her imposing marble desk, smiled from below the eyes, and nodded toward a large chair the staff referred to as 'the hot seat.'

I sat.

"I think you'll agree that I've always been supportive of you in the past."

"Oh, absolutely," I gushed. "I've never felt unsupported or left out to dry, in fact..."

"Satanism," she interrupted me, "Satanism," repeated, "is what I'm hearing from parents." She held up a letter, much like Joe McCarthy during the Army-McCarthy hearings, waving it as if she were fanning herself with smelling salts. "They claim you're teaching Satanism."

"You mean it's not part of my curriculum?" I tried to humorously deflect the charge.

The Dowager Empress was not amused. Above all, she did not like surprises, a trait she shared with most school administrators. I thrived on them, a trait I shared with many teachers.

In all fairness, she had covered for me in the past, from charges of Desertion of Phonics to Dereliction of Duties when I had lost a kid or two on a camping trip. But she obviously took this charge seriously. There was a very vocal parent group set on aligning our district's curriculum with the Holy Scripture and deleting all books from the school library containing references to human body parts below the waist.

"They claim that you're teaching chanting, meditation, Yoga, and worship of Nature gods."

Well, it was true. I did all those things. "It depends, I guess, on your interpretation. We do choral reading and response. I guess that's chanting." (I didn't mention our Oming exercises). "I have them clear their minds to open up their imaginations for creative writing. That's your meditation." (I didn't mention the hushed, sacred visualization practices). "We relax and stretch, that's the Yoga." (I didn't mention our daily ritual of performing the Obeisance to the Sun). "And you know I do a lot of nature study and environmental sciences." (I didn't mention playing panpipes to rouse the Spirit of Pan and other Nature Sprites).

"Well, that sounds innocuous enough. But be cautious. You know how I hate surprises." She dismissed me with an adamantine glance over the rim of her glasses.

I smiled, nodded vigorously, profusely thanking the Dowager Empress for her personal and professional support, and beat a hasty retreat.

Boy, I thought, I'm glad she didn't mention our dramatization of the sex life of the salmon. Just the recollection of Matt wriggling and writhing about on the stage as he squeezed dozens of ping pong balls out of his pants leg, gave me pause.

But not for long. It was time to get back to class, time to worship the sun and roust the great god Pan from his slumber.

Chapter 12
The Russians are Coming, The Russians are Coming

On a crystal clear October morning in the late '50's, the slightest hint of Autumn tinged the air. America was smug and secure in its post-war bed dreaming of the World Series, (the Milwaukee Braves {Spahn and Sain and pray for rain}) defeated the hated New York Yankees, 4-3). America was Fifties self-satisfied, comfortable in its position as leader of the world. Without warning, disaster struck! We were wrenched from our Leave It To Beaver and Father Knows Best reveries with the soul-chilling news that the Russians had launched a satellite, which was even then speeding overhead, violating the infinite airspace of the U.S., and we were thrust once again into war. The Cold War, the Nuclear Arms Race, the Space Race, closed its bony grip on our national hearts, and determined all of our destinies.

It was October 4, 1957. The Soviet Union had successfully launched Sputnik I, the world's first artificial satellite. About the size of a football and weighing only 183 pounds, it loomed enormous from our planet-clinging vantage. The hated Russians had won the first encounter of the New Age, the Space Age, in which the first nation to gain dominance in space would rule the world. From its fortress on the Moon or an orbiting space station, the new planetary Dominatrix could hold the world hostage with nuclear weapons, city destroying lasers, or other 'weapons of mass destruction'. And the damn Russians were winning!

November 3, the Russians struck again with the launch of Sputnik II, carrying Laika, the first space dog. This victory was followed by the knockout punch, the orbiting of a human, Lt. Yuri Gagarin. We were on the ropes. The first three rounds won decisively by the Russians. National humiliation turned to steely political resolve. We would win the race to the moon, dammit, or risk becoming Russian pawns, stumbling through our days in a mindless fog, calling each other comrade, and wearing ill-fitting, unbecoming gray suits, rumpled and hopelessly out of fashion.

How could this have happened? How could mere Slavs (Life magazine reported that 60% of Americans thought the term was Slob rather than Slav) whip the most technologically advanced nation in the history of the world? We didn't even know who or where they were (50% of Americans misidentified the Soviet Union and The United States on a map of the world, assuming, perhaps, that the U.S. just had to be larger).

With the resolve that won the war, we turned our attention towards regaining supremacy in the air and in the classroom. We launched a national initiative to produce the world's finest missiles and the world's most astute scientists and mathematicians. America called in the big guns: enter infamous Nazi scientist, Dr. Wernher von Braun. Wernher had gained notoriety by devising the V1 and V2 missiles that rained anonymous death on England during the war. For an in-depth biography of the famed doctor, refer to Tom Lehrer's song:

Don't say that he's hypocritical,
Say rather that he's apolitical.
"Once the rockets are up, who cares where they come down?
That's not my department," says Wernher von Braun.

With the space program in good Nazi hands, the Federal government, for the first time, turned its attention toward education, previously the sole property of state and local governments. Pursuing a national imperative to achieve an 'educational moonshot', schools were mandated to produce scientists, and quickly.

In spring of 1959, I participated in a scene replicated all over the country by bright young scholars eager to join their nation's war on ignorance.

I sat at an old oak desk bolted to the floor in a classroom in Chicago, taking a national test to see if I would join the ranks of future scientists of America, the bright-eyed students who would help our humiliated nation regain its proper place in the world. I passed, and in fall of 1959 twenty-five of us entered a large Chicago high school, the second wave of the new generation. We belonged to an elite corps referred to simply as 'The Hundreds Program'. We few, we proud, we overtested, were largely isolated from the rest of the student body, kept full day in classes designed just for us.

Oh yes, and we did reach the moon first.

Chapter 13
It's All Upstream

W hy oh why oh why O, why did I ever leave OHIO?" I sang out in my blustery baritone. It was my theme song, sung whenever I found myself stuck in a ludicrous position. Sung often. Like now, as I stood back from our class' nine and one-half foot salmon, wedged immovably in the classroom doorway.

Why had we built an absurdly massive paper maché fish? Why had I not measured the door? Why, oh why, oh why O?

"Mr. K, Mr. K, it's stuck," grunted Jeremy, stating the obvious, himself wedged beneath the monster. "It's not gonna go," wailed fragile Belinda, near tears. "Not gonna work," darkly observed Tim, the class pessimist. "Let's try it backwards," suggested Morgan. "Cut off the fins," offered Sonya, the class realist. "We've almost got it," urged Nary, the class optimist.

I had taught long enough to know that the kids were always right. Since, however, they all disagreed with each other, this knowledge was of scant help. "Okay, okay, let's just take a break and think this through." The kids were glad of the relief from pressure, except Nary, who, all forty pounds of her, kept trying to drive the stuck salmon upstream. "Nary, give it a rest."

"Okay, Mr. K., but I think we're almost there."

"I'm sure we are, Nar, but let's take a breather."

I gave Jeremy a hand, pulling him out from beneath the behemoth, and soon we were all standing inside the classroom surveying the implausibility of the task at hand. The gargantuan fish was wedged behind the gills and before the pectoral fin. From the rear it looked as if a St. Bernard were attempting to mate with a Chihuahua. Impossible.

Lost weekends, late nights working on the framework for this monstrosity flashed before my eyes. Standing back, I looked around the room for an adult, hoping against hope that there was someone other than me in charge. All I saw was a sea of little eyes looking up at me, waiting for a wise proclamation.

I regrouped my rather meager personal resources and sighed, "Okay, the fins have to go."

"All of them?" reluctantly asked Nary.

"All of them. Get the hacksaws."

The kids scrambled to the tool box, wrestled briefly for control of the weapons of mass destruction, and fell on the beast, sawing away at the dorsal and pectoral fins, till they hung, dangling, the pvc struts severed.

"Okay, now heave." And with a right good will we fell to the rear of the gargantuan fish until "Pop," it sprung free, carrying me and six kids sprawling out into the hall.

"Free at last, free at last, thank God Almighty it's free at last."

We fell eagerly on the mangled carcass and like an army of ants hauling off a cricket, we hoisted the almost ten foot fish and carted it down the hallway, out the doors, and into the bed of a pickup truck I had commandeered from a fellow teacher. Although the dilapidated truck lacked tags and plates, I reasoned that a huge salmon was so obvious no cop would think of pulling us over. And, after lashing, off to the Natural History Museum we drove, kids trailing in a van.

Having been alerted by the museum staff, a photographer from the local paper greeted us and chronicled the transfer of the oversized salmonid from the truckbed, through the museum doors (yes, it fit), and safely ensconced in the display window, adorned with appropriate signage, the class artwork, models, and research. We reaffixed the severed fins and fell back to admire our work.

The newspaper reporter approached me. "So, Mr...Mr. Kuttner, is it? How did you manage this enormous task?"

The kids interrupted, "Oh, Mr. K didn't have anything to do with it. We did it all ourselves."

"How cool is that?" I thought. Now that's success.

Chapter 14
Testing and Dinosaurs
Compare and Contrast

Standardized testing is the gasoline combustion engine that drives the educational system. It is large, smelly, noisy, wracked by violent explosions, and relies on dead dinosaurs (in the form of old ideas) for its fuel. It also develops an irresistible momentum. Gas engines and testing are both obscene profit-makers for megacorporations, businesses that are committed to defending those profits to the last breath, businesses willing to go to any lengths to maintain their stranglehold on the public. Corporations willing to rape, pillage, and lay bare the natural and cultural landscapes in their quest for power, fueled by greed.

Both testing and oil employ a huge battery of highly compensated lobbyists who buy and sell politicians like bricks of fly-covered lard in a corrupt, polluted marketplace. The resulting handcrafted legislation ensures no change in policy, undercuts any chance for real reform: either renewable energy sources or gentle, intriguing authentic assessment.

Allow me to retreat to the earlier days of testing, before it was so deeply entrenched in the American psyche as essential and immutable. One hundred years ago testing was embraced as a quick and painless way to separate the wheat (the leaders) from the chaff (the followers). Those dedicated to making the American educational system more efficient argued that only 6% of the boys would become professionals, and only 7% of the girls would become teachers (about the only 'status' job open to them). "The rest would work as servants, cooks, waitresses, laundresses, saleswomen." Their schooling would prepare them specifically for these tasks and not waste time on liberal education. A job analysis would be done on, say, waitressing. The specific waitressing job skills would be the curriculum. The chute into which one was herded at the educational slaughterhouse would be determined by (of course) IQ tests.

Sociologists, such as Charles Ellwood of the University of Missouri, argued that all the children in the state serve 'life sentences' in public schools until they could pass an exam demonstrating proficiency. Until that time they would not be allowed to marry or have children. The true dream of these 'pre-fascists' was to have testing control breeding (actually control the racial stock of the country). Ellwood's 'breeding' exam was the precursor of today's California State High School Exit Exam. We've passed a law to keep the kids in school until they pass the Exit Exam (or age), we just haven't figured out how to keep them from breeding.

Thus, through exceptional management and organization, we have completed the circuit of making schools more efficient, delivering a product that neatly fits our market needs. If, after all, the 'free' marketplace demands bigger, meatier breasts, why bother to raise chickens who can walk? If the marketplace demands compliant workers, content to perform impersonal, repetitive tasks ad nauseum, without question or complaint, why encourage students toward critical thinking? Walking (and thinking) would become irrelevant. Impediments, actually, to a smooth functioning, efficient, waste-free industrial world.

Considerable ethical considerations aside, it simply doesn't work. There has never been a scintilla of evidence that standardized tests improve learning. Amidst the deafening national clamor for tests and more tests, no one has ever shown that testing helps kids learn, nor that they make bad schools good.

This however has not slowed the progress of the testing industry. As of 10 years ago, 148 standardized achievement tests had been published in this country for elementary students alone. Only 34 of them have gone out of print!

And the stakes have been raised. The tests, many of them produced by textbook manufacturers, drive the curriculum. The tests decide what will be taught. In New York City, Joel Klein, the new hired gun Superintendent has offered up to $40,000 apiece to superintendents who raise test scores. At my school, we received a memo from our Superintendent with a printout of last year's test scores. Those scores below the 50th percentile were highlighted in radioactive, dayglo yellow. The memo ordered us to create lessons that addressed all the 'low score' areas, and turn in a schedule of times those areas would be addressed each day, so that, "I can schedule my day to drop in classrooms and see those test items being taught."

You think people don't teach to the tests? Think again.

Chapter 15
Samurai Table Day

"K raaaaaang!" One long, uninterrupted bell signaled an earthquake drill. "Under the tables, grab hold of a leg, a table leg" (I added, remembering the time Matt had grabbed hold of a girl's leg, while she howled, held it firmly, until I had pried his fingers off, and wrapped them around the leg of the table. "But I was just following orders," he grinned mischievously). "Hold the table leg and be quiet. Listen for instructions or the all-clear."

"Mrff, mrff." A muffled grumble.

"Quiet, please." Imperiously.

"Mrff, mrff."

"What *is* the problem?" I asked, from my teacher position, braced in the doorway.

"Nick's stuck," Kristen ventured, breaking the school imposed gag rule.

"Oh, man," I thought, "This is really it." I left my official teacher-in-the-portal earthquake posture and strode over to check out the situation. Sure enough, there was Nick, his buttocks and legs anyway, kicking and squirming, his torso and head wedged rather securely under the table, looking like a bloated tick trying to wriggle in for a solid purchase, or like Pooh Bear caught inexorably in Rabbit's hole.

I placed my hands on his butt cheeks to further wedge him, just as quickly removed my hands, thinking, not for the first time, "What am I doing?" I then grasped him by his hind legs, leaned back (for Nick was a large child and stuck) and dragged him out, speaking reassuringly, "That's okay, Nick, let go. It's okay. It'll be alright. Go hide under my desk."

Once released from the considerable pressure of the table on his head, he looked up at me pitifully, gamely smiled, and, breathing heavily, scuttled off to gain his new position, clacking across the floor like a large beetle.

"Klaaaaaang!" The all-clear sounded suspiciously like the earthquake bell. But, if the class had been sufficiently quiet to hear the first bell, one could reasonable sequence them and realize it was safe to come out from cover. If, however, as sometimes happened, the class missed the first bell, you were then out of sequence with the rest of the school, ducking, holding, and covering while the remainder of the school went securely about with the rest of their business. Upon these occasions we would often remain huddled under our tables until word reached us from the outside world, often by complete happenstance when someone in the know strolled by our classroom, surveyed the scene, and asked us, "What in the name of Pete are you doing?"

"Practicing," I'd often say, "It pays to practice. Can't hurt."

But this time it had hurt, not physically (other than a slight reddening of his upper back, Nick seemed sound) but emotionally. So why had Nick become stuck? Let me take you back a few months.

The day before school had started, the Imperial

Administration had canceled a combination 4th and 5th grade class, swelling the numbers of the existing classes. My own class had grown to 32 fourth graders, a rather unseemly number, and one which outnumbered the available seating in my room. I brought in a card table and folding chairs, but that was obviously a stopgap measure. There was simply no room to ramble about the class, which was my preferred mode of movement.

I made a mock-up of my classroom at home, wracking my creativity to find more space, asking myself what was truly essential. I determined that chairs take up a minimum of four square feet, statically, and eight square feet, dynamically (room to pull back and push in). That redounded to a stunning 33% of my classroom! They had to go.

The next day, dubbed "Samurai Table Day", we stacked all the chairs in the hallway, sawed the legs off the tables, made pillows, and ensconced ourselves on the floor, there to remain until the end of the year. The kids and I got such a kick out of it, I was to repeat this performance year after year. "Mr. K," the new class would ask, "Is it Samurai Table Day yet?"

And what of Nick? Even though he couldn't fit under the lowered tables (low clearance ahead), he reveled in his newfound status of Under the Teacher Desk Monitor. Even when there was no earthquake drill, he'd often be found reading, writing, or just hanging out under my desk. More than once, I'd sit and be surprised by his big, round, cheerful face mooning up between my legs, smiling up at me, "What's up, Mr. K?" Occasionally, while seated at my desk, I'd hear a crunch or feel my shoe gummed up by one of Nick's snacks, stashed in his secret place.

Scraping Kraft Cheese Whiz off my shoe, I'd think, "A small price to pay."

Chapter 16
A Call For Standards

If I may, I'd like to throw my bull in the ring, dump on the bandwagon, circle my wagons, grasp the unicorn by the horn, and add my modest voice to the midnight choir croaking for new standards. Tough new standards with heightened accountability, replete with dulcet incentives, and pungent disincentives.

These standards, however, are not of the Trivial Pursuit-Jeopardy variety, those minibytes of information known as "factoids" in the old science framework. These are standards to live by, standards on which to build a culture, a workable society.

What might such standards look like? Well, first of all they'd be standards that all citizens could attain, otherwise it'd just be more of the same elitist multi-tiered tracking to which we've become accustomed over the checkered course of Western history. (The partially clad Gandhi was once asked by patronizing Western reporters what he thought of Western Civilization. He replied, "I think it would be a good idea.")

From the onset, the Ten Commandments were crafted

to ensure failure. Sure, there can be injunctions against theft and murder, but when you start to include "thought crimes" such as worshiping and idolatry and especially coveting, one ensures that every mortal falls short of the lofty goal, forced to live a life of repentance and guilt, doomed to perdition.

How exactly does one measure coveting? Is there a standardized test to determine when harmless admiration crosses the murky line to the sin of the covet? Can the government make the distinction between a gentle appreciation of my neighbor's ass, and a true coveting of his or her ass? Can I hold my neighbor's ass in esteem? In high esteem? At what height does esteem of his or her ass become an honest to goodness covet of his or her ass?

There is a definite injunction against me coveting my neighbor's ass for myself, but is it okay to covet his or her ass for a friend? ("I know it would be unforgivable and sinful for me to covet Joe's ass. Would you mind coveting his ass for me? And any time you want some ass coveted, just let me know.") Could I covet ass for another neighbor? If we all coveted each other's asses for each other, it seems to me we'd be on the road toward creating a truly generous, sharing society.

But no, that's not really the point, is it? The point of Western education is not, never was, to create a generous, sharing society, a society (gasp!) of equals. The Obscenely Wealthy White Old Men who structured our educational system, who invented "intelligence" testing to weed out the infirm and unfit, who crafted the "performance assessments" to ensure that half the kids fail (guess who?), who throw out test items that too many students answer correctly, who craft standards that are unreachable, irrelevant, immaterial, and hopelessly out of touch with the reality of our kids, (take a breath, this is a long sentence), these Obscenely Wealthy White Old Men (OWWOM) have devoted their lives to ensuring a multi-tiered society of enslaved disenfranchisees (largely people of color who may or may not be enraged, no great matter), of mindless, compliant, unquestioning workers (may be any color or class, middle or below, the essential trait being an unwillingness and/or inability to make a decision or take a stand, other than what color or how sexy the next purchase will be), and Them, the OWWOM and, of course, their trophy brides.

Standards. This is a call for standards, organic standards, and a rejection of standards imposed by "higher" authority. Any standard worth achieving is one that grows from the community that it purports to measure. The standards must arise from the ashes of our reelingly classist, racist culture. Standards must finally be our opportunity to define ourselves, to project our aspirations. Who do we, as individuals and as a collective, wish to become? How do we want to define our relationships with each other and the environment? And then, only then, can we establish means of achieving these goals, of helping each other achieve them. Surely it is obvious that the method of bell-curving our culture into winners and losers, haves and have nots, dreamers and zombies, is a tragic failure. We can achieve nothing ourselves unless we achieve for all.

Hey, maybe that's a start, the First Standard. We, the citizens of planet Earth, do recognize the value, the humanity, the dream rights, of ourselves and of each other. We hereby reject the "Old Standard" of divide and conquer, of being played against each other for the profit and the power and the pleasure of the few. You may no longer measure us against each other, for we are all of equal worth.

It's a start.

Chapter 17
Surprise Party

M r. Kuttner, they want you in the office."
Shit! Now what had I done? Mrs. Carey, parent and
Mama Bear extraordinaire, looked pleased as she presented
me with the information. She had a few reasons to be
displeased with me, but I thought we had worked them out.

It was Mrs. Carey who, at Back-to-School Night in the
fall, had publicly challenged my credentials to teach her
son, Kyle. "Your program sounds okay for the faster kids,
but my son's always had a reading problem. What are you
going to do for him?"

It had been only a year since I had announced at First
Grade Back-to-School Night that "the acquisition of
reading is shrouded in Mystery. It's rather magical,
actually." Parents' pleasant social smiles had frozen in a
rictus of fear and disbelief. Concern had washed over the
assembled parents like a charging wave crashing on a flat,
defenseless beach, drawing all the comfort out of the room,
and pulling it back into the sea like so much flotsam.

I was now teaching fourth grade, but word of my

"rather magical" belief in acquiring reading and other skills had obviously swept around our school.

"And," Mrs. Carey added, sensing advantage, "I notice that your outline doesn't mention missions. Are the kids going to make missions? That's what I remember from fourth grade. Making missions out of sugar cubes."

"No," I admitted, "We're not going to do missions. I feel that the Mission Period in California represents a real low point in our moral and ethical history. We'll address the issues, but I don't want to glorify the period." Parents were now squirming in their seats. Engaging in one of my quick internal dialogues, I pleaded with myself to stop, to pull out before I dug myself such a pit I'd never see the sun again, never feel the warmth of summer, never view another friendly face.

But, lacking the intelligence God gave the turnip, I added, "I don't do Concentration Camps, and I don't do Missions."

"And besides," I ventured, hopelessly trying to provide an escape route, "We've got a terrible ant problem." This last reference to sugar cube missions was so far removed from the original suggestion, and so otherwise lame, that I realized I had completely lost the crowd. I could almost hear the parents speaking to themselves, "He's daft. He's a nut. God, I hope he's not dangerous."

As the parents filed out, shell-shocked, depressed at the serious deterioration of the California Public Schools since their childhoods, I fell to damage control. I felt we had had a good year. Sure there were some miscues. I had been horsing around with some of the kids, chasing them around the hallway. Kyle had grabbed a piece of candy and attempted to escape. I had brought him down from behind, sitting on top of him and pinning his arm underneath as I dislodged the candy from his fist. The loud "Pop" and his howl of pain informed me that I had dislocated his shoulder. Kyle (yes, the same Mrs. Carey's son) was the star pitcher for the Little League team, a major position in a small town such as ours. The title game was the following day. Very bad move!

But I had apologized and we had moved on, or so I thought.

"I'll watch the class as you go down to the office," Mrs. Carey had smiled at me. Did I detect a hint of smugness, perhaps even retribution? Oh, well, nothing to do now.

"So the Boss wants to see me?" I inquired of Barb, the two-headed dog guarding Hell's Gate (the Boss's Secretary).

"Nope, not this time," she shared, not bothering to look up. As I walked back down the hallway, befuddled, I heard some rather loud squawks coming from my room. Oh, no! Running now, I threw open my door.

"Surprise!" the class of thirty fourth graders shouted, as Mrs. Carey beamingly thrust a large white layer cake at me, the words, "Happy Birthday Mr. K" scrawled in blue cursive icing across the top. Kids were scattered in various poses around the room, some under tables, some jumping out of the coatroom. Kyle, and his good buddies Matt and Brandon, were positioned on the tabletops, saxophones hanging around their necks, all grinning hugely. They began to belt out a tune. Could have been "Happy Birthday" or any one of a number of unrecognizable songs. No one seemed to care.

I glanced back to Mrs. Carey, smiling broadly, eyes and all. "They really do love you, you know."

And at that moment, I did. I did know.

Chapter 18
"One Size Fits Few"
A Look at Educational Standards

There has been a society-wide hue and cry for standards, especially high standards, in American education. 'Hue and cry'? I had to look that one up. I generally think of a hue as a tint, or coloration. Turns out that 'hue and cry' derives from Old English for 'pursuit of a felon accompanied by loud outcries'. This is telling. Education, of course, being the felon pursued by a loud, angry, outcrying mob of citizens. Appropriate.

Standards are viewed by most as an expectation, a goal by which to measure progress. It has, however, become a demand, a blunt, complex set of demands accompanied by exhortation and threat, encouragement and intimidation. For hand in hand with standards comes standardization. The view that 'one size fits all'. Now, in our marketplace view of education as output, that makes sense. One can readily see that in a world where 24 squared off (cubic) tomatoes fit in a box, you don't want a box with 25. Do you sell it for more or give a tomato away free? It slows down the process of tomato transfer, you see.

In the marketplace tomato world, we have been brainwashed to accept the corporate view that what really matters is the size of the tomato, that and the shippability. Tomatoes have been bred to be of uniform size and consistency so that efficient harvest machines can replace workers to increase production. Never mind that modern tomatoes never ripen and taste like Jello boxes. Never mind that modern tomatoes can withstand the assault of an entire match of tennis and still bounce. Never mind that they are heartless and soulless and smell like cellophane. What matters is that they look alike, they survive shipping, and they fit. The mesmerized public continues to buy them because, on some primordial level, they remind us of tomatoes. We have bought the devil's bargain that ensures increased corporate profits and has stripped our culture of real, wholesome experiences, full of truth and beauty.

I can accept that in a world of mass production of titanium bolts and condoms, you want some uniformity. We want aircraft bolts and nuts to match threads. Although, with regard to condoms, one would really have to challenge the notion that one size fits all.

But in the world of education, standardization only makes sense if you persist in viewing people as floppy discs, subjected to software that 'writes' data on the receptive discs. These discs are then subjected to high pressure (Stuff-it) to increase the amount of data stored, and released from pressure (Expand-it) to retrieve the data.

Think what kind of hell would be unleashed on the Electronic World (henceforth known as eworld) if these pliant little discs would begin rejecting data implants because 'I'm not interested', or 'I'm not sufficiently motivated', or 'This makes no sense in my world', or 'Dad hit mom last night and the cops came to get him and I'm really not very focused on this test item right now', or 'What am I going to do with this?', or 'I didn't have any breakfast', or 'My dog got hit by a car', or 'Everyone tells me I'm stupid', or 'These zits are ruining my life', or 'Hello? There's a human being in here.'

We have so far removed market tomatoes from true tomato-ness, that they won't even reproduce. They are so weakened by corporatizing, that they really no longer exist outside of the market realm. They have lost their vigor, their flavor, their color, their blemishes, their idiosyncrasies, their individuality, their uniqueness.

And we dare threaten to streamline, to standardize, to corporatize students in the same way. Under the phony banner of accountability, of standards, of standardization, we threaten to strip our students, the fruit of our culture, of their vigor, their flavor, their blemishes, their individuality, their humanity.

We're losing the tomato battle, but the human battle is one we can ill afford to lose. Our true challenge is not to shape our students to fit the box, but to allow them to shape themselves into creative, critical, thinking beings, into individualists, blemishes and all. The real problem with standards is not that they aim too high, but rather that they aim too low.

Chapter 19
Public School, Private Pain

His new, plaid shirt looked more than pressed, starched, probably. Although it was a warm early September morning, the first day of fourth grade, the top button was secured, so tightly as to cause a slight puffiness of his neck. His lips were level, neither a smile nor a grimace, but assuredly a sign of uncertainty. Set comfortably apart in his smooth, almost beautiful olive face, his deep, warm eyes, the color of chestnuts, peered out appraisingly. He held his little body stiffly, awkwardly, weight set evenly on both feet, almost as if expecting to dodge a blow.

Although I greeted him warmly, profusely, even, and posed for a photo with him and his little sister before the white board, enjoining all who entered to: "Welcome to 4th grade. Don't worry, do the best you can, then be happy", it was clear that he was not following the imprecation; he was most certainly worrying.

I felt such a pang in my chest that I wondered if I could start the day, if I could regain the bounce in my step, the cheeriness in my voice and manner. Each time I looked at

Rio, my son, on his first day in public school, I wondered if I had made the right choice, right for him, that is. Did my need to be with him, spend more time nurturing him, override his best interests? Should I have left him, shy but happy, at his safe, little private school?

His mother and I had divorced when he was in kindergarten, his sister in pre-school. Although we had both remarried within two years, all literature suggested that the kids, even when they liked, even loved the stepparents, harbored a slim hope that the parents would rejoin. Made up bizarre scenarios in which both steps, out for a friendly stroll, were devoured by mountain lions, fell off a cliff, developed leprosy, anything that would allow the natural parents to fall into each others arms, correcting past mistakes, leading perfect "Parenting Magazine" lives.

As I look back at that time, a bit apprehensively, I can't remember how long until he unbuttoned that top button, relaxed a bit. Was it weeks, months? I don't recall. Did he ever relax? I don't really know.

We, he, had our moments. He dressed as Jim Thorpe, world's greatest athlete, storied victim of the racism, classism of the time, and delivered a life story in the first person. His full size sculpture of himself as the golfer he was to become accompanied his autobiography. His poetic voice in his Just-So stories, in his rambling Mary Sloin series (the woman who squatted down on a milk pail and delivered a child in the stables, smiling all the while) was his and his alone.

A very sweet, kindhearted boy, he was quite sought after by the girls. When we did Big Friendly Giant puppet shows, Rio appeared in no fewer than six different performances. The fact that his puppet was a waiter, replete with tray and towel, made him quite versatile, but more than that, his choice of puppet gave a deeper insight into his self-image. A waiter. One who waits. One who waits on others. One who serves, rather than is served.

In the fourth grade Olympics, he was the fastest kid in the grade. First of ninety. A naturally talented athlete, full of fluidity and grace, he was a welcome addition to playground, and, later, organized sports. The ultimate team player.

He was a math whiz, always in the top group, always ready with the answer, but only if called on. I rarely remember him ever offering. He was uncomfortable on his own, shining, standing out. Talented as he was, he much preferred the background, the shadows. He organized a quartet of boys to sing at the talent show, accompanied by me on the guitar. What was the song? Blowing in the Wind? If I Had a Hammer? some song of protest and brotherhood. He was not shy in a group.

But, years later, he reported to me how he had anguished, how he had wandered, lonely and alone, at recesses, sobbing softly to himself. Why had I not interceded, he wondered, accusingly? How could I allow this kind of heartache to be visited upon my son, whom I adored? How? Why?

The clues were there. His always tentative nature. His reluctance to initiate. His passivity, which I confused with natural pacifism. And, of course, always his stuttering. Mild at times, extremely severe at other times. He'll outgrow it, I projected. It's not him, just something he does. He's popular enough that it won't be an impediment.

The clues were there, I simply didn't pick up on them for a variety of reasons. I truly believed that public school provided a more wide-ranging education, including, especially, social adaptation. His cloistered private school represented a small social gene pool, drying up year by year as the 'healthier' kids were transferred out to public schooling, until only the real misfits were left, those who could not possibly survive the rigors of the natural world. I had simply wanted to shelter him, to protect him.

But he suffered. And he suffered alone. And although there are reasons and excuses and rationales and time behind us, it still hurts. It hurts both him and me.

And to think, I, who could so freely dispense advice, wise observations and admonishments about others' children, I could be so blind and ignorant about my own. It's humbling.

Chapter 20
Where Do the Mermaids Stand?
More Standards

Robert Fulghum in his coy tome, "Everything I Need to Know I Learned in Kindergarten", reports an important incident from his teaching. He had gathered a rather rambunctious crew of kids in a church basement and proceeded to teach them a game called Giants, Wizards, and Dwarfs. There was no real purpose to this game. It did not, as we are currently wont to say, address any higher educational standards, nor did he demonstrate any "best teaching practices". It was however meant to be fun and to have the children interact and wear themselves out.

When Fulghum cried out "Giants, Wizards or Dwarfs", the children were to run around in a wild frenzy and group themselves, find like-minded individuals and huddle together, awaiting further instruction. Amidst the bedlam, he felt a tug at his pants leg. Looking down, he saw a small, wide-eyed girl, tugging and looking up at him. She asked him, "Where do the mermaids stand?"

Fulghum responded, teacher-like and authoritatively, "There are no mermaids, there are only giants, wizards, and dwarfs. Find your group and stand with it."

She persisted. "There are mermaids. You see, I am one."

This little girl refused to give up her dignity or her identity. She did not identify with any of the available options and refused to leave the game and go stand where the losers stood. She intended to participate without loss of self.

Fulghum asks: "Well, where DO the Mermaids stand? All the "Mermaids"--all those who are different, who do not fit the norm and who do not accept the available boxes and pigeonholes?

Answer that question and you can build a school, a nation, or a world on it."

So the little Mermaid stood by Fulghum, not as a loser, but as an equal.

"(They) stood there hand in hand, reviewing the troops of Wizards and Giants and Dwarfs as they roiled by in wild disarray.

It is not true, by the way, that mermaids do not exist. I know at least one personally. I have held her hand."

****

Like it or not (personally, I do) the early world of children is full of young mermaids. It's packed with kids who are different, whose reality is not reflected in any district, state, or national standards. Kids who cannot accurately be measured by officials whose minds are sealed shut, or threatened, by difference. Kids who cannot accurately be measured at all. Call it homophobia, racism, classism, genderism or whatever ism you wish, it is really a fear of, a lack of compassion for, difference. Its true name is intolerance, fueled by an absence of respect and ignorance. We are experiencing a national epidemic of intolerance, of xenophobia, fear of that which is strange. Xenophobes pack the White House, haunt the hallways of the House and the Senate, man the military, preside over corporate board meetings, make the rules. And the offspring of their intolerance is a set of rules and standards that determine who will succeed and who will fail.

How do we prepare kids for success as citizens? How do we prepare them to fulfill themselves? At my local high school, kids who exercised their legal rights to exemption from the STAR test were pressured and badgered to conform. Kids who exercised their First Amendment rights not to participate in the school sanctioned 9/11 assembly were threatened. Kids who attempted to exercise citizenship by attending a peace rally and teach-in were detained and punished. A friend of mine called her brother, a high school principal in Sacramento, and asked him what he would have done. He acknowledged that there was a liability issue in kids leaving campus (although they leave it every day for lunch). His solution? He said he would have gone with them, solving the liability dilemma. Now, that's leadership.

What a great question: Where do the mermaids stand? One of the truly important questions in education. Where do we stand: those of us, those of our charges, who are different, who are not standardized? Can we really refer to our educational system as ethical if it does not adequately address that question? I think not.

Chapter 21
Pumpkin, Pumpkin, Roll Along

Charlotte Moose, Nazi bus driver, was not in a good mood. Let me put this in perspective: given that a good mood for Charlotte resembled a simmering, slightly pre-psychotic rage, a bad Moose mood was not a pretty sight. In fact, teachers and children had been rumored to be turned to stone (or at least sticks of chalk) merely for regarding her in one of her ugly, Medussaic incarnations.

The Empress Dowager, our mutual boss and Grande Dame, had just informed her that due to parent complaints, Moose could no longer wear her anti-environmental protest cap reading, "Save a Tree, Wipe Your Ass with a Spotted Owl," while she drove the school bus. The parental protest had not been over the environmental ethics expressed, but the use of the 'a' word.

Moose, so blunt that she had been called 'dull', didn't get it. "I don't see what the f**k's the problem," she observed.

Charlotte was not in a good mood. Usually this was no great concern to us. We simply poured into the storm cellar until the tornado had passed, emerged, assessed the damage, and went back to work. We, however, had a field trip scheduled for that day. So I warned the kids that, "The Moose is on the Loose", and, in rather subdued fashion, we filed mutely onto the bus for our annual trip to the Pumpkin Patch.

Now, as I'm sure most of you know, kids and dogs have a collective memory about as long as a mayfly's, or perhaps an inanimate object such as a wave, so it wasn't very long before someone (it wasn't me, honest) broke into song. In fact, we had been singing pumpkinny songs in class, so it required little incentive for the whole group to pick up:

"Pumpkin, pumpkin, roll along.
I like tamarinds, I do.
I like pork and rice, I do.
While I sing this song."

"That's it. Shut it!" Bellowing like an enraged elephant seal, Moose blasted over the bus loudspeaker. As redundant as it was for her to use a loudspeaker, she savored its impact. "Raise your right hands. All of you!" (she glowered at me in the rearview). My hand shot up as well.

"Now place your raised hand over your mouth. Now! And let me not hear a peep, not a peep" (she challenged) "till the end of the trip. No speaking for *any* reason. None."

Whew! Not even the smart asses were tempted to ask what she meant by that. A pall settled over the bus, depressed and silent, caught in the vice-like clutches of fear, just the way Charlotte Moose liked it.

As is often the case in hostage situations, time dragged slowly by. Sixty of us sat in leaden silence, staring glumly out the window at a now gloomy landscape. Eventually we approached our turnoff for the Pumpkin Patch. Much to my mute, mouth-covered glee, however, Moose showed no signs of slowing down for the turn. My spirits soared as she raced past the exit, hurtling blindly onward for another ten or fifteen minutes until she caught my eye in the rearview. "Mr. Kuttner, where's the darn turnoff?"

"Mmrf, mmrf," I squeezed out through fingers tightly pressed against my mouth.

"Did we pass it?" her mouth pursed, forcing the words out.

"Mmrf, mmrf," I nodded my head.

"KUTTNER!" This time we all jumped. She had used the loudspeaker.

Summoning what little courage I possessed, I leisurely removed my hand and, in as controlled a manner as possible, calmly informed her, "We passed it about ten miles back."

"Why the hell didn't you tell me?"

"I tried to," I pled as innocently as I dared, "but all I could say was 'mmrf, mmrf'", I mmrfed as pitifully as I could.

Her eyes narrowed dangerously, her crow's feet grew to crevasses as titters and light giggles spread throughout the bus.

I was thankful I was in the back seat, far from the reach of her ham hock hands. I glanced around for emergency exits I could use rather that venturing near her seat when it was time to debark.

Enraged though she might be, Charlotte Moose was no fool. She chose to cut her losses and raced to the next town to turn and retrace the route. Although it felt to us that she could drive on her own fumes, the mood in the bus had lifted to sedate mirth and the gentle optimism that arises when certain death or disfigurement is narrowly averted.

And although it took a long time for Charlotte Moose to warm to me, she never again asked me to cover my mouth.

Chapter 22
When General Bullmoose Leads,
the Leaders Will Follow

A s Charlie Wilson, CEO of General Motors and Secretary of Defense (formerly the Department of War) testified to the Senate, "What's good for General Motors is good for the country." This led cartoonist Al Capp to create General Bullmoose as the epitome of the ruthless capitalist. General Bullmoose used to bellow, "What's good for General Bullmoose is good for the USA." So what's good for the General Bullmooses, and where are they leading us? They're leading us in a holy crusade for Standards.

The Standards Movement was born of a publication titled, "A Nation At Risk" in the early 1980's. It gave rise to a number of responses, notably King George the First's National Education Summit in 1989. At this gathering, goals were set for the year 2000, including "being first in the world in science and math by the year 2000," (sounds a lot like the Educational Moonshot of the Sixties, doesn't it?). This libidinous alliance of government and business produced a superabundance of government agencies to "remedy the educational malpractice" of the nation.

A much abbreviated list of these agencies/commissions includes: the National Education Goals Panel (NEGP), the Secretary's Commission on Achieving Necessary Skills (SCANS), the National Center on Education and the Economy (NCEE), the New Standards Project (NSP), the Learning Research and Development Center (LRDC), the Mid-Continent Regional Educational Laboratory (McREL), the National Education Standards and Assessment Council (NESAC), the Business Coalition for Educational Excellence (BCEE) and the National Council on Educational Standards and Testing (NCEST).

The unfortunately acronymed iNCEST did not keep other outfits from their own malapropisms such as the CLAP (California Learning Assessment Program), named after a venereal disease. Staffers changed the name to CLAS (California Learning and Assessment System) after the absurdity of naming the state test after a Sexually Transmitted Disease was brought to their attention. I thought it totally appropriate.

The education summit also birthed the National Assessment Governing Board (NAGB), a clique of politicians, administrators, and businessmen who, over 3 martini lunches (low sodium olives), chased by non-fat mocha lattés long pull double espressos (no foam), set policy for the National Assessment of Educational Progress (NAEP), "The Nation's Report Card." This was chaired by Darvin Winick, President of Texas Bio (you don't want to know what they do), and Ed Donley, Chairman of Air Products and Chemicals (air products?).

The Second National Educational Summit of 1996 was a cozy affair attended by a handpicked group of 40 conservative educators and businessmen. No students and only one or two teachers were invited. At this affair, the silver-tongued Republican Governor of Nevada, Bob Miller, complained, "Too often we seem too willing to accept underachieving standards for a Beavis or a Butthead."

At this point, the corporate leaders outed themselves from their closet, left the director's chair, and took center stage. They formed a front group called, Achieve, Inc., which then hosted the next two summits, the National Education Summit of 1999, and the Education Summit of 2001, co-hosted by IBM CEO Louis Gerstner. This brazen cooption and corporatizing of education is led by the Board of Achieve which includes 3 Republicans, 3 Democrats and 6 CEO's: Louis Gerstner, CEO IBM, Phil Condit, CEO Boeing, Keith Bailey, CEO Williams, Craig Barrett, CEO Intel, Ed Rust, CEO State Farm, and Art Ryan, CEO Prudential Financial.

It should come as no great surprise that Kurt Landgraf, CEO of ETS, Educational Testing Service, which brings you the SAT exams, as well as CAHSEE, the high school exit exam, the AP exams, and a Testing Database with 20,000 different tests, is a staunch supporter of Achieve, Inc.

IBM and Gerstner, along with a supporting piratical band of corporate thieves, have also hosted the European eLearning Summit and the Latin American Basic Education Summit, to coerce other nations to adopt our own educational standards based on world domination. Our motto: Make the World Safe for American Businesses (in reality, multinational corporations), In other words, exploit willing workers in countries without the power to withstand economic subservience and environmental degradation.

It beggars the imagination. Not only do the Captains of Industry have the audacity to proclaim what American kids need to know, they have the unparalleled hubris to proclaim it for the entire planet.

What's good for General Bullmoose is good for the world, if you know what's good for you.

Chapter 23
Free Time

Big Brian usually moved as if weighted down, struggling through a medium denser than air, as if every motion were a titanic battle. He sweated, he huffed, he grunted, gave every indication he'd really rather not. Big Brian did not like to move.

While planning our fall camping trip, I had assigned Brian his own personal assistant to prod him along the trails I had chosen for us. Always, I drove the kids, and drove them hard when we hiked, reasoning that dawdling opened the door to mischief and mayhem.

I had relaxed from my earlier days of camping with kids. I still produced an agenda, mostly for the parents, but no longer was it noted to the minute:

9:05 First Rotation: Tribe 1 at Point A, Tribe 2 at Point B, etc.

9:25 Warning whistle

9:28 Second Rotation: Tribe 1 to Point B, Tribe 2 to Point C, etc.

9:30 Begin Second Rotation

It was offensive, but harmless. Most parents were impressed and it relieved stress amongst some of my co-workers who were, it was rumored, ill at ease at the prospect of camping with me, Staff room Anarchist, Closet Beatnik.

But I had, as I said, relaxed from the rigor of earlier schedules. I did still drive them on the trails, stopping just short of actual cattle prods or other devices currently out of favor in California Public School classrooms.

"In the good old days," Rafe Slye used to say, "We could whack the little tikes to keep them in line."

"Yeah," I'd respond, "Now you've got long division."

Rafe was notorious for torturing kids with obscenely immense long division problems, always holding the threat just above their heads like an inane Sword of Damocles. Rafe had coached pole vault at our school in the days when kids could high jump higher than they could pole vault. I always suspected there was some tenuous connection between that and his refusal to use calculators, but I never quite pinned it down.

I assigned Big Rodney to Big Brian as his personal prod, confident they could appreciate their mutual statures, a match super-sized in heaven. Big Rodney was Little Rodney's father, an imposing Native American Highway Patrol Officer, fearsome behind his reflector shades. Privately, I knew that Big Rodney was about as tough as the Pillsbury Dough Boy, but I figured his cop status and frightening demeanor was sufficient to motivate Big Brian to keep on the trail and keep trudging.

"Okay, troops, fall in," I bellowed. "Head 'em up, move 'em out," I sang out, beginning my brisk pace. After about forty-five minutes, we reached the trailhead leading down to the beach.

"Officer Rodney," I caught Big Rod's eye, "We'll meet

you at the bottom." Big Rodney understood that Brian would not be able to keep up with us as we climbed down the hundreds of stairs to the ocean. Chuffunk, chuffunk, fifty or so feet started the long descent at a safe but steady clip. At the bottom, I assembled my charges on the sand and set the boundaries, "No one even near the water without an adult. Ever. You may climb up the dune" (a very high, sandy bluff faced the ocean) "to that line of vegetation" (I pointed up the slope). "No further. First violation rates a timeout. Any questions?"

"Yeah, why's the sky blue?" asked Daniel.

The kids issued a collective groan under the weight of their impatience to break loose.

"Good question, Dr. D. We'll chat later. Okay, get rolling."

Half the kids broke for the dune, half for the water, having already forgotten the warning about adult supervision. I whistled them to a halt, caught up, and walked them down to the surf. For the next hour kids alternated between rolling down the dune and rolling around in the surf.

Big Brian showed up not too long after we had hit the beach. Tireless, for once, he scaled the dune, rolled, careened recklessly down the sand like a jackknifed big rig, wiping out smaller kids, billowed to a spattered stop, scrambled to his feet and back up the slope again.

Twenty minutes of dune play and over to the ocean. Gleeful and carefree as a seal, Big Brian flopped into the surf as a wave rolled in, picked him up and deposited his beach ball of a body up the wave slope. As the surf receded, Brian rolled back down the wave slope, mixing with shells and pebbles, until the next wave picked up and carried him up the slope again.

Big Brian repeated this operation, smiling hugely,

cycling back and forth from ocean to dune, several times, building up a thicker and thicker layer of encrusted sand and grit on his clothes, till he looked like a colossal corn dog, or a massive decorator crab, bizarre and ungainly.

On one return to the sea, Big Brian rolled down to my feet and lodged there, crusted and bulky. Lying at my feet in the receding surf, he looked up at me, smiling all the while, and asked, "Mr. K., when are we going to have free time?"

For once I was speechless.

Chapter 24
Baboons, Penis Envy, & More Standards

Susan Ohanian, in her provocative little diatribe, "Standards, Plain English, and the Ugly Duckling," says, "Male baboons exchange greetings by yanking on each other's penis. I don't know how Fortune 500 CEOs, media pundits, and politicians greet each other, but I do know that only 1% of their DNA differs from that of baboons and that 99% of what they say about public education is hooey."

Lest baboons suffer by the comparison, Ohanian continues. "And if one hundred baboons sat in a computer lab, they'd produce *Hamlet* sooner than one hundred CEOs would tell the truth about the relationship of their advocacy of one-size-fits-all educational standards to downsizing, outsizing their salary packages, the sideswiping of middle- and working-class America, and the subsidization of private education at the expense of public."

One problem with many CEOs (the most power encrusted) is that, in addition to being unconscionable bullies, they are also liars and (surprise, surprise) hypocrites to boot. As American corporations relocate ("outsize" is their euphemism) abroad to countries with willing, underpaid, ununionized workforces and few environmental restrictions, they claim that American education has failed to provide the skilled workers to protect business from foreign competition.

The CEO of AT&T, Robert Allen, tripled his package to $20 million a year, even as he laid off 40,000 employees. Disney's Michael Eisner, in the midst of Disney lay-offs, received $203 million in one year alone. His predecessor, Michael Ovitz, received a severance package of close to $100 million, to celebrate his unsatisfactory performance. How much do you think the fired Disney workers got?

Al Dunlap, CEO of Scott Paper, laid off 20% of his workforce, then collected a cool $100 million by selling his trimmed down, efficient company to Kimberly-Clark. Louis Gerstner, the same jerk who decries the low standards of American education, pocketed $21 million when he moved from RJR Nabisco to IBM. Then, adding injury to insult, "At the same time he was co-authoring a book berating teachers for not producing a sufficient supply of world-class workers, Gerstner fired 90,000 of IBM's 270,000 workforce, those highly trained technical experts he says schools aren't producing."

This does not even address the fabricating of books and the corporate sleight of hand to simulate business growth. In a parliamentary government, George Bush and Dick Cheney would have been humiliated by the revelation of their collusive, insider trading theft, and been asked to resign. However, as General Bullmoose said, "Don't do anything crooked, unless it's legal."

These CEOs, these Captains of Industry, these blocks, these stones, these worse than senseless things, these hard hearts, these cruel men of wealth, care nothing about anyone or anything other than their own compensation packages and the veneer of profitability of their corporations. As average CEO compensation jumped 23%, 3.26 million American workers were fired. Did the workers benefit from the streamlining of industry, from the slimming down, from the "downsizing"? I think not. And as long as shareholders hold profit above decency and humanity, these thugs will continue to write their own laws, write their own standards, and terrorize the citizenry.

These corporate raiders, despicable as they are, are merely symptoms of a deeper ill. They are pustules on the body politic, fed by the overwhelming desire for unearned wealth in this country. For some reason, many of my generation, the Yuppified Baby Boomers, so worship money they are willing to sacrifice whatever scruples they may still possess in the name of profit. So long as we, the stockholders, demand profit at any cost, can we really be surprised that there are scoundrels willing to please us and advantage themselves?

And these overpaid callboys blame hardworking students and overworked teachers for a weakened economy? The very words "Standards and Accountability" should turn to ashes in their mouths. We need some ethical standards and moral accountability applied to industry, first. Maybe then we'll listen as they talk about educational standards and accountability.

Chapter 25
Sita and the Flag

Early in 1991, at the height of the First Gulf War, guns and tirades blazing, the flag disappeared. The flag, which I hyperbolically referred to as, "The World's Largest American Flag," was not, in fact, the world's largest American flag, or even, perhaps, California's largest American flag, but the damn thing was huge. Barn size, billboard size, which of course it was: a billboard for patriotism. A public declaration that the owner was a barn sized patriot accompanied by the implied statement/question: I'm a huge patriot, are you?

But one Monday morning, the flag flew no more. At first I didn't notice its absence as I drove to work. Odd that I had been so incensed when Ben Blue, proud owner of the "World's Largest American Flag" had despoiled the landscape, co-opted the communal airspace, invaded the 'commons' and hoisted the flag, blocking my view of the beach and the ocean. Odd that I had been so incensed and yet, when the intrusion had been retracted, when the weapon withdrawn from the wound (been stolen) and the landscape, the viewscape, returned to 'normal', I hadn't noticed. I knew something was different, as if an old friend had finally shaved off that ridiculous mustache and you couldn't figure out what was different, even though you had begged him to shave for years.

The flag had been stolen again, and Ben Blue was furious, outraged, almost incendiary when I saw him that day. His son, Dick, was in my class, and I had scrupulously avoided discussing flags, pennants, banners, or even large pieces of fabric with him, anything that might remind Ben of his barn-sized patriotism and his evangelic zeal in displaying that attitude. When he picked up Dick after school, he was on a self-righteous tear.

"Hey, Mr. K., Mr. K., I'm gonna have a surprise for you, yes I will, a surprise for you." To Ben, anything worth saying was worth repeating ad nauseum. "Yes, a surprise."

I assumed he was referring to our walking field trip down the road the next day, to his farm and soil supply store. Ben had made a very handy living out of scraping home and building sites, removing the surface soil to his processing plant, screening the till for rocks, sorting the rocks into various grades of aggregate and then reselling everything, often to the same folks whose property he had scraped in the first place.

He sold gravel for driveways, larger grades for concrete mixes, and the largest grade of aggregate for landscaping. The remaining soil was sold as topsoil to replenish the sites Ben had despoiled.

We were to visit Ben Blue's Landscaping to select rocks to border our new pond at school. Dick Blue, Ben's son, was particularly excited about Ben's plan to fill a loader with rocks and have Dick, all of 9 years old, operate the loader, drive it a mile down city streets back to school, and dump them at them pond site. Needless to say, I had not notified our insurance carriers of Ben's plan.

I assumed Ben's plan had something to do with the

loader or forklift or whatever in God's name he was planning on allowing his son to pilot. But the next morning as we approached Ben's Boulder Shop, Ben came bustling up to us breathlessly, "Mr. K, Mr. K, O boy, O boy, you're going to love it, boy, you're going to love it."

I held up my hands, palms towards him, in an appeal for calm. "Hi, Ben, what's up?"

"O boy, Mr. K, I've got an extra, I saved an extra flag, and we're gonna raise it and I was saving it for you, and, o boy, the kids are going to get such a kick out of it..."

My heart sank, shoulders slumped. Jesus, not another flag. Not just another flag, but The Flag.

"Come on, kids, come on, climb up on the truck. We're gonna raise the flag. You'll love it. The flag," he repeated, gushing.

Scrambling, scuttling over each other, the kids climbed aboard his flat bed truck, eager to handle the flag, hoping, perhaps, that some of the magic, some of the power, some of the barn-sized patriotism might rub off on them.

One child, however, hung back, didn't clamber aboard the Patriot Express, didn't even look at the truck, the flag, or the massive staff, a commercial boom poised to hoist and fly the flag. It was my daughter, Sita. She appeared to be intently focused on a rock, an undistinguished bluish chunk of chert.

"Sita," I whispered at her. "Seets", she looked up at me furtively, caught my eye, and vigorously shook her head "No" to my unasked question. I gave her a small smile, just the corner of my mouth. She small-smiled back, silently communicating her deep distaste at the prospect of handling the World's Largest American Flag.

I sighed contentedly, my heart swelled with pride, and I turned back to oversee the swarming mass of patriots positioned for the hoist, poised for glory.

Chapter 26
Make The Pie Higher!

Despite a few tens of thousands of votes that were uncounted, mangled, spindled, ignored, buried, packed and shipped on a slow boat to Haiti, we succeeded in electing a man of education, a man self-promoted as the "education candidate". The "education candidate" has become the "Education President." George Dubya claims to have presided over "real educational reform" during his tenure in Texas, the "Texas Miracle." And what was that, George?

Under Bush, Texas' high school completion rate was 46 out of the 50 states (US Dept. of Ed). Teen drug use rose 30% (Texas Commission on Alcohol and Drug Abuse). George, perhaps still hung over from his drunken, drug-hazed years at Yale, feigned chagrin. Although Bush spent $10 million on abstinence only education, Texas has the fifth highest teen birth rate (US Dept of HHS). Under Bush, Texas teacher salaries dropped from 36th to 38th in the nation; their total salary-benefit package was 50th (NEA).

Under Bush, Texas had the second highest number of children living in poverty, and it held the dubious distinction of housing the highest percentage of children without health insurance (US Census Bureau). His state ranks 48th in spending for health per capita (US Dept of HHS).

But under George, the "compassionate conservative", test scores in Texas did rise. The Carnegie Report states, however, that the disparity between white and minority students increased and schools exempted up to 50% of their students from the tests, in order to artificially inflate the results. Sounds a bit like the "creative accounting" of Bush's closest financial friends and advisors, such as Enron's "Kenny Boy" Lay, to give the appearance of educational health and solvency.

But who better than George himself, to sound the clarion call for higher standards and accountability? I give you, in his own words, George Walker Bush, the "Education President."

(Washington Post writer Richard Thompson, in honor of National Poetry Month, took the liberty of aesthetically arranging actual quotes of George Bush II in poetic form):

Make the Pie Higher
by, George Bush

I think we all agree, the past is over.
This is still a dangerous world.
It's a world of madmen and uncertainty
and potential mental losses.

Rarely is the question asked,
"Is our children learning?
Will the highways of the Internet

become more few?
How many hands have I shaked?"

They misunderestimate me.
I am a pitbull on the pantleg of opportunity.
I know that the human being and the fish
can coexist.
Families is where our nation finds hope,
where our wings take dream.

Put food on your family!
Knock down the tollbooth!
Vulcanize society!
Make the pie higher!
Make the pie higher!

This is what our leader, the unchallenged head of the Free World, the Scourge of the Infidel, the self-appointed Steward of the Standards and Accountability Movement, the iconographer of educational reform, demands of us. Dare we naysay him? Can we resist being swept along on his charismatic pantleg of opportunity? No, I say. Wherever he follows, we must lead.

But, as my pappy was fond of cautioning, "If you're going to follow a horse's ass, you'd better carry a broom and a shovel."

Chapter 27
Above Reproach

His thin, drawn eyebrows arched in frozen disdain over milky, rheumy blue eyes, now hidden behind wraparound, bombadier-style shades. His pale, creamy flesh, the consistency of old school paste left out to set up and crack, looked vaguely like a head from Madame Tussaud's wax gallery gone bad, set aside to be recycled and put to better use. Dangling jowls only accentuated the sneer permanently carved into his tight, bloodless lips. The overall effect was one of absolute absence of generosity. If, as Abe Lincoln had said, "After forty, you have the face you deserve," Edge Rodmann must have done a lot of damage to a lot of life forms to have earned such a ravaged visage, devoid of life, hope, or warmth.

In the all too typical, perverted realm of public school hierarchy, he had, of course, risen to the office of school superintendent.

"Son of a bitch. Goddamn it, Kuttner," the lifeless face erupted in fury, small red lines spreading across his cheeks, as if a windshield had shattered in slow motion, "Damn it," he shouted, "Everything was fine until you showed up. You always screw everything up."

Although Rodmann was screaming, thrusting his bleak face at me menacingly, I focused on the veins and arteries pulsing and throbbing on the sides of his neck, fascinated by the physical manifestations of apoplexy.

It was several days before the start of school. The class lists had just been posted and, as usual, someone had monkeyed with them over the course of the summer so they no longer reflected the balanced classes we teachers had painstakingly created before summer break. Several teachers and I had gone in to ask 'The Edge' what had happened. In his predictably unpredictable way, he had exploded at me, gone postal.

From years of finely tuned experience talking folks down from bad drug trips or disarming irate parents bent on destruction, I began to croon to him, soothingly as one might speak to a colicky child. "Now, Edge, no one's saying that you had anything to do with it." Mellow and slowly as if I were disarming a bomb. In fact, I had often considered donning protective gear whenever I went to see him about a problem. "Nor even," I continued, "That it was intentional, we're just concerned about the make up of the classes."

"Goddamn it, Kuttner," Rodmann's needle was stuck in a groove, unable to process input.

"Well, Edge, we didn't mean to upset you. Maybe this is not a good time. We'll talk later." And we attempted to beat a graceful retreat.

Once the clear and present danger had subsided, I began to, as we say in California, 'feel my upset.'

"Damn it, I'm sick of it. I'm going to file a grievance. That shit has got to learn he can't bully us like that."

Katy, ever moderate, perkily suggested, "Whenever I

have a problem with him, I go in and tell him directly. He's much better one on one."

Against my better judgment, I agreed and walked right into his office. "Edge," I began to speak as I crossed the threshold, before I lost my resolve, "Edge, you may not speak to me or anyone on staff that way."

Rodmann fixed his porcine, little eyes on me and launched into it. "First of all, you may address me as Dr. Rodmann" (Edge had 'earned' a doctorate from a mail order house, much as I had 'earned' a doctor of divinity degree from the Universal Life Church, although I didn't demand that people address me as 'Monsignor'). "You know what's wrong with you, Kuttner?"

"I'm sure you're about to tell me."

"You think you're above reproach."

"What does that mean, Edge, that I don't accept you as my parent and me as a scolded child? You've got that right."

"See, you're above reproach." The next twenty minutes were filled with Edge Rodmann's small-minded invective, liberally punctuated with 'you', various expletives, and multiple finger pointings.

As is often the case with doomed prey, my mind quieted and drifted to kinder, gentler scenes, in my case, a leisurely river trip on a smooth stretch between rapids. I remembered nothing of Rodmann's petty attack other than the phrase, oft repeated, "Above reproach."

Upon leaving his office, I went to a local shop and ordered a coffee mug with "Above Reproach' in gothic gilt lettering.

As often as I dared, I'd approach Rodmann, cup extended, lettering towards him, and inquire sweetly, "Coffee, Edge?"

Chapter 28
Life on the Planet Bizarro

I awoke this morning somewhere else. Everything looked the same: the placement of furniture, the filaments of the ceiling's spider webs in the same state of disrepair, the familiar shadows, the everpresent valley fog with its diaphanous dispersal of the rising rays of the sun. Everything looked the same, but everything was different. There could be no mistaking it. The surface remained, but the core substance had shifted.

I didn't doubt for a moment but that I had slipped into an alternative reality. This I knew, but which one? I immediately dismissed the Planet Gor, the shadowy counterpart of Earth that mirrors Earth's orbit, trailing along in the penumbra, sensed by some, but unobservable to most. No, Gor was a medieval twin of Earth. There would be no pillowtop mattress, no bedside phone.

But if not Gor, where? The question nettled me, like a large bug on the collar, felt but just out of sight. I opened up the morning paper (another hint I wasn't on Gor) and glanced at the headlines. "President unveils No Child Left

Behind Act, challenges schools to perform. Promises to accept no excuses. Announces new Get Tough policy".

Of course, I had awakened on the planet Bizarro! Now, for my younger readers, Bizarro had been the invention of Dell Comics, the creators of Superman, Supergirl, and Superdog. It was intended to highlight the differences between the straight and narrow America of the Fifties, and the potential degradation associated with fraternizing with the evil empire of the Soviet Union. It was meant to be, I think, a pretty strange amalgam of communism, reefer madness, and rock and roll. It represented the modern Limbo to which those who fell from the Grace of the Protestant Ethic were consigned.

The planet Bizarro served as Earth's not-quite-evil twin. Not truly evil, but seriously messed up. On the planet Bizarro everything that was revered and commonplace in the middle America of the Fifties was disrespected and slightly tweaked. Bizarro had its own Superman and its own institutions that had somehow run amuck. Superman and his friends were not Barbie and Ken beautiful, but ugly and rough around the edges. Superman didn't (gasp) shave regularly. The coiffures and the lawns were unkempt. The houses of suburbanity had dilapidated fences and the paint was peeling off the walls. The citizenry was routinely late to work and the trains didn't run on time. In fact, no one seemed to care about time, or about work, for that matter.

Businesses did not run a profit, in fact, they seemed to just give things away. Kids didn't pay attention in school and displayed a lack of respect for authority. In short, everything went to hell. War was Peace, Peace was War. Nothing made sense.

Yes, I was sure of it. I had awakened on the Planet Bizarro. A world where the United States, and only the United States, could claim the right of "First Strike Deterrence", could retain the right to Nuclear Checkmate those with whom we disagreed.

A citizenry who had elected George Bush ("favorite books: Norman Vincent Peale's The Power of Positive Thinking and Billy Graham's The Last Crusade) as the "Education President". A citizenry who allowed the CEO of IBM to write the educational standards for its youth. Yep, the Planet Bizarro for sure.

I know, I know, there has been some groundwork laid. We have dipped into Bizarrohood in the past. California elected soft shoe dancer George Murphy to the Senate (as Tom Lehrer said, "Now we have a Senator who can really give us a song and dance"). We also foisted Tricky Dick on the American Scene, author of the best-selling memoirs, "I Am Not A Crook, and other Lies and Misdemeanors." And let's not forget the failed actor, the Great Communicator Ron (seen one redwood, seen 'em all) Reagan, in bed by 8 every night ("Good night, Mummy") who could read someone else's speech as soothingly as Milk of Amnesia. We have also been the scene of Arnold's "The Terminator" launching a totally buff assault on the Governor's Mansion. Life imitates very bad Art.

But with the elevation of George W. to the lofty post of "Education President" we have convincingly raised the Bizarro bar, we've really "made the pie higher". You may think that his appointment of his business friends as the watchdogs of the corporate community is bizarre. You may think that his appointment of ecophobic businessmen as watchdogs of the environment is bizarre. You may think that his use of Dick, "I never saw a stock option I didn't like", Cheney as ethics czar is bizarre. But all of it pales against the gauzy backdrop of George as Education President.

Yes, this morning I woke up on the Planet Bizarro, and it's not funny.

Chapter 29
The Fledglings

argarita was a bird. She didn't play at being a bird; she was one. A truly lovely little being, incapable of deception, she never twittered or tweeted, never flapped her arms and jumped off lunch tables. She certainly didn't preen. Naturally and unselfconsciously beautiful, Margarita's dark olive skin, clear open intelligent eyes, and lustrous ebony hair needed no preening. No, Margarita didn't play at being a bird; she was a bird.

So was Lupita.

One classroom, two birds. It was a first for me.

I had had werepeople, shape shifters, in class before. Monica had been my first. She had been a wolf, a small one, playful but leery.

"Mr. K, Monica's under the table again."

"Monica, honey, it's time to come out, now."

"No," she barked, regarding me warily with huge, unblinking saucer eyes. Monica never blinked. "I can't, it's the moon," she explained without further elucidation, her eyes glinting with caution.

Small and meek though she was, there was an air that hovered about her that caused me to be wary of putting my hand under the table, of reaching for her or moving too abruptly.

"Mr. K, she's creeping me out," complained Shannon, 'creeped out' by anything out of the ordinary.

"Shannon, it's not really your concern." I tried to nip the complaints before they spread around the classroom. "When you need to shift your shape, you can have the space," I offered.

Shannon recoiled, wrinkling her nose, "Oh, yuck!"

Several kids looked interested in taking me up on the offer, so I scurried to change the subject before every boy in the room dove under their tables.

Margarita and Lupita had come to their avian spirits, as far as I knew, independent of each other. Although they had known each other for years, both spoke Spanish as a first language, they did not conspire in avianity. It was a set of realities that was non-conspiratorial and natural. That made the fluttery nature of the classroom all the more appealing.

Neither of the girls ever used their avian beingness as an excuse. As difficult as it must have been to secure a pen or pencil by flight pressure or grasped by talon, they never complained. They'd just get quieter than usual, and focus more intently.

At P.E. they'd participate as they could, but they ran with an odd gait, like the flap-glide of a finch or woodpecker, never quite landing with their full weight. Whenever it was windy, however, both would usually face the wind, with one or more body part in constant motion. More than once, on extremely windy days, we'd find them at the far end of the field, downwind, and I'd send the 'P.E. Monitors' to escort them back to the class.

Our classroom faces out on the garden. Often, when the

weather was good, or even passable, I'd suggest, "How's about some garden time?"

The kids almost always responded enthusiastically. It was a time for the girls (my 'fine feathered friends') to shine. Flitting from bush to bush, from flower to flower, they were totally at home. Never did I see them so at ease, so relaxed. Occasionally I'd catch sight of them out of the corner of my eye, pluck off a bug or two from a leaf and pop them into their mouths. It'd happen so fast, at first I wasn't certain it had happened at all, but the spark in the girls' eyes, their eye shine, convinced me it was so.

The last day of the year I hug my kids on the way out the door. Margarita and Lupita dawdled until all had gone, then came up to me together for the last hug. I bent over to hear their super soft "thank you's" and watched them walk slowly, side by side, almost hesitantly out of the class, out of the nest. They stopped, briefly shook themselves, looked back over their shoulders, smiled shyly, and launched themselves, soaring elegantly off into the world. Fully fledged.

Chapter 30
No Child Left Behind
No Child Left Unrecruited

Sharon Shea-Keneally, principal of a high school in Vermont, was surprised when she received a letter from military recruiters demanding a list of all her students: names, addresses, and phone numbers. She, as is typical, invites the military to share in career days and fairs, but keeps student information confidential, as do most schools. Shea-Keneally says, "We don't give out a list of names of our kids to anybody. Not to colleges, churches, employers--nobody."

But when she asked for an explanation for the demand, she was cited the No Child Left Behind Act. Within the labyrinthian depth of the act, is a provision requiring secondary schools to provide military recruiters with access to the school grounds, and also access to all sensitive contact information for each and every child. Provide that, or risk a cutoff of all federal aid for education. All aid.

The American Association of School Administrators

feels that this is "a clear departure from the letter and the spirit of the current student privacy laws. It's a slippery slope. (We) don't want student directories sent to Verizon either, just because they claim that all kids need a cell phone to be safe."

The new law gives students the right to withhold their records, but school officials are given great latitude in the implementation of the law. Many schools are just handing over the records to recruiters without informing kids, parents, or anybody, leaving those most affected without any say in the matter. What do you think your area schools are doing? Why not ask them?

The President of the San Francisco Board of Education feels that, "the privacy implications of this law are profound. For (them) to ignore the concerns for the privacy of millions of high school students is not a good thing."

The new law undercuts the authority of some local school districts, such as San Francisco and Portland, that banned recruiters from school contact on the grounds that the military is anti-gay and discriminatory. These districts will now allow contact, but educate kids as to their legal rights to withhold information.

Although military recruitment has exceeded its recruitment goals for the past two years, despite being barred from some schools, recruiters feel driven to explore all avenues toward impressment of children, and the training of them to take the lives of others. Military recruiters claim that they will aggressively follow up on student contacts, including personal contacts, even if parents object. "The only thing that will get us to stop contacting the family is if they call their congressman," says Major Johannes Paraan, chief of recruitment for the North East quadrant of the US. "Or maybe if the kid died, we'll take them off our list," he said.

No way, ever, would I allow that soulless ghoul access to my children's files. What say the administrators at your local high schools? Why not give them a ring-a-ding-a-ding? Call the school board while you're at it.

Public schools also underwrite the culture of militarism by nurturing and subsidizing classes for JROTC, kind of a junior rangers club (Geez, Ranger Rick, can I really take sniper lessons?), but with real weapons. Some school boards think it's a mite bizarre to offer classes in marksmanship and weapons training (over half of JROTC units do this) concurrent with a drive to safer more secure schools. At the same time schools are busy prosecuting kids for bringing guns to school (zero tolerance), they are being paid by the feds to offer gun classes in school. Go figure.

Vets for Peace, the Albany Friends Meeting, and the Catholic Worker mounted a successful challenge to JROTC in their town. "Military programs promote obedience rather than leadership. They foster a culture of war and violence rather than peace. We prefer that young people get credits in serious academic subjects."

Depressed and increasingly desperate, we wander across this desolate and barren landscape looking for direction, searching for an ethical signpost. What we get is military training and a politically expedient call to arms for high standards and accountability. But what we really want is a sign that someone truly cares.

Chapter 31
Going With the Flow

Into the river I plunged. Caught up by the eddies, the bottom and bank rollers, lifted by the tidal bore, borne along on a whirring sea of wheels. I felt I could raise my hands from the handlebars, close my eyes, fling my head back, and sing in safety, cradled by mass, embraced by the flow, in complete security. Ahhh!

> *Do you know how it feels?*
> *Do you know how it feels?*
> *Spinning your wheels,*
> *Under thighs of steel,*
> *Do you know how it feels?*

Ahh, the ease, the restful ease. I am a migrating goose in formation, carried effortlessly along by the wingbeats of my fellows. I am a dust mote suspended in the comfort of Brownian Motion, weaving, wefting, gliding. I am a slithery Silver Salmon, massed for migration, pectoral to pectoral, oozing downstream, easy and seamless.

I am a Western Sandpiper, flock-flashing in the sun, swooping and soaring in unison...perfection. I am a drone, a worker, caught in the primal hum of the bees' mantra. Buzz, buzz, busy, busy. One heart, one love.

I am...I am... **ALONE!** Where the hell did everyone go? One moment I am one with the wheels, one with the pulsing mass, one with the All. The next I'm the *only* one. Alone. Where the hell did they go?

****

It had begun like every other dreary Tuesday: chill, damp, overcast. And too damn early. Not for the first time I thought, "Guy, you are my hero. You detest rising at this obscene hour, this chrono absurdity, yet you do it. Repeatedly. Have I ever told me I'm my hero?" I growled in my gruff early morning basso.

Rolling out of bed, literally, my body too stiff to sit up like a normal human being, I give in to gravity, caress it, and am caressed in turn, sinking out of bed to the floor like an old slinky, kinked and disjointed. Several minutes of groaning stretches follow, prelude to grappling to my feet and waddling to the doorway, listing back and forth as if on a drastically pitching deck. As I reach out, groping the wall for balance, stumbling to the door on the path to coffee, my gateway drug, I think again, honestly awed, "Holy, shit, what is this going to be like in twenty years? In ten years, for God's sake? How the hell do people do it?"

Stage 1: Cup in hand, not yet fortified by the coffee, but greatly reassured by the habit, the tradition, I renavigate the familiar route to bed, and settle back under the covers to complete the routine. Why do ruts have such a bad press, I wonder? Sure it's hard to hop out of them, but who wants to? Without my morning ruts to keep me grooved, who knows what the hell would happen? Unabashedly, I stand for ruts, I decide. Who would have thunk it? "I'd like to order the morning routine, with a side of irony, please."

Stage 2: stare at the coffee, stare at the clock, stare at the coffee, stare at the clock. So far my morning has been completely routinized. Stretches, two cups of coffee, a satisfactory evacuation of the lower intestines and all other regions south of the Equator. Kiss my wife. Tell her she's beautiful. Kiss the dog. Tell her she's beautiful. Let one out, keep the other in bed. Look back at the clock.

Stage 3: Oh, my God, decision time. Have I got enough time to bike to work? Factors in play: con: too old to bike, pro: too old not to, con: too late, pro: not if I get my ass in gear, con: out of shape, pro: only way to get in shape, con: it'll feel bad, pro: it'll feel worse if I don't.

Stage 4: Okay, this is really decision time. That was pre-decision. Today, I bike, and damn the consequences.

****

Always the same route, always. It's not the easiest, in fact it's probably got more hills than any other, but it is the shortest and the quietest. I had seen notice in the paper that there was to be a Mass Transit today, a community demonstration of the positive aspects of bicycling, referred to as Critical Mass. But I had forgotten. I'm not some young hippie or college student, I've got a real live job, for God's sake, and I've got to get there on time. No time this morning to demonstrate with a bunch of moonbeamers.

But I round the corner, and there they are, and there I am, and they forgive me who I am, and what I am, and they open their collective arms, wings, fins, whatever, and embrace me. And I am caught up in the moment, only for a moment, but at that moment, that's all there is. Caught up in the moment. Ahhh! Suddenly I'm forty years younger, no job, no schedule, no place I'm going to, no place I've been. Ahhh!

I am not an aging Anarchist fighting workplace Fascism. I am not an old man in a young woman's profession. I am simply a piece of the organism, integral and insignificant, both, going with the flow.

Is there any other way to go?

Chapter 32
No Child Left Behind
No School Left Unthreatened

Several years ago, Santa Bush brought the American public a special Christmas gift. No, not the possibility of a world at war in perpetuity, the gift that keeps on giving (and keeps taking away). No, not that little thing. On January 2, 2002, George "The Education President" Bush signed into law the extension of the Elementary and Secondary Education Act, commonly known as the "No Child Left Behind Act."

Catchy title, don't you think? Obviously not written by the staffer who wrote the "We're gonna hunt 'em down, and then we're gonna smoke 'em out, and then we're gonna flush 'em out, and then we're gonna kill 'em" speech. "No Child Left Behind." Can't really argue with that. So what is it?

Analysts don't really agree on its implications. Hell, they don't even agree on the length, running from 670 pages to 1100. Maybe the higher estimate had the DC phone book stuck to the back. Anyway, there are some things I can tell you about it since, if you have kids, are going to have kids, are a kid, were a kid, pay taxes, or know someone who pays taxes, it applies to you.

It requires all states to have a system of annual testing in place, using 2001-2002 scores as the baseline.

It requires all students to be tested, and tested in English, including non-English speakers and all students in Special Education.

All data will be disaggregated, which means that results for kids from subgroups such as economically disadvantaged (poor), all ethnic groups, disabled, and limited English proficiency will be segregated from the group scores and also be reported separately.

All states will publish annual report cards detailing how their kids, including those individual groups, scored on the standardized tests.

All new teachers are to be "highly qualified", certified in the specific curriculum areas in which they teach. Title 1 teacher's aides must have a minimum of an Associates Degree or its equivalent (same meagre pay).

Extra fed funds are to be used to set up "scientifically based" reading programs (for scientifically based, substitute "phonics based").

Penalties for "school failure" will become more draconian each year. Schools reported as "failing" (low standardized test scores) must offer students the option of transferring to a "high performing" school (look out Beverly Hills High, here they come). After three years "low" schools must offer to pay for tutoring for the kids, after four years they must pay to transport kids to schools of their choice. They may then be "restructured" by the state/feds.

The impact of this legislation on wealthy suburban schools should be minimal. Inner city urban schools and rural schools will bear the brunt. From the perspective of rural schools, the testing issue becomes even more absurd in that every person educated on the issue of standardized testing has found that the smaller the sample, the more unreliable the results. And I mean everybody. There is no one in this country, short of a few ignorant and ill-informed politicians and businessmen, who think that a small test sample shows anything at all. The motives, then, must be political rather than educational.

No one thinks it's a good idea to hire unqualified teachers, but many wonder whether the feds are the ones to determine who's got the chops and who don't. It also means that small schools that have one teacher teaching many subjects, are going to be censured.

The issue of classifying schools on the basis of their test scores is, of course, highly questionable. That aside, one wonders how to increase a school's "performance" when "higher performing" kids are shunted off to "better" schools? And then to require small, underfunded, rural schools to pay to ship kids around the county is an issue that frightens everyone. Since schools will never be able to afford that, can you imagine how much pressure will be brought to bear on teachers and kids to jack up their test scores? The fiscal solvency of entire districts will be at stake.

But of great interest to me are the hidden provisions of the No Child Left Behind Act. One of them requires that all secondary schools in this country send to military recruiters a list of all their students: names, addresses and phone numbers, making public schools a de facto arm of the military.

Given that the federal military recruiting budget was

increased to 3 billion dollars a year, it wouldn't appear that schools would be needed to augment the military's activities in this area, but Don Rumsfeld and Ron Paige, co-authors of a threatening letter to school officials, obviously disagree.

Chapter 33
One Nation Under God

H ey, Mr. K. how come we never say the Pledge?"
"Hey, Mr. K., did we forget something?"
"Hey, Mr. K., did we forget the pledge?"

If I had a nickel for every time some little guy had brought that to my attention, and had invested those nickels wisely, say in $24 worth of glass beads, I'd have a tidy retirement sum.

My 'forgetfulness' in the realm of ultra patriotism is legendary. My responses over the years have varied from the pungent, "Oh, yeah, right," to, "So many things to do, so little time," to the teacher/parent perennial favorite, "Later," or "Did I?"

One year I used the ploy that we didn't have a flag in the room. This didn't foil my group at all. Seaton and Claymore met at home that afternoon, fervently pursuing a project. The next morning they greeted me, as pleased with themselves as Betsy Ross, and ceremoniously presented a miniature version of the American flag, 'stitched' together with hot glue. I, of course, greeted the gift with exclamations of joy, tacked it to the wall, and proceeded to forget to conduct the pledge whenever I could. I was to encounter these two little ultra patriots over a decade later as they stood on the sidelines of a peace march, hurling mindless invective at the silent marchers as we roiled by. The unvarnished truth is that I have hated the Pledge of Allegiance ever since the infamous American Nazi, Joe McCarthy, bullied the Federal Government into including the phrase, "under God", into the pledge in the mid-fifties, to distinguish our sacred, God-fearing democracy from the Godlessness of the Soviet Union. I grew up during the early days of the Civil Rights Movement, disgusted by the blatantly undemocratic Apartheid under which this country operated. As a public draft resister during the early days of the Viet Nam War, I vowed not to support racist and militarist policies. And certainly never, ever to utter that mindless drivel, the dreaded Pledge of Allegiance.

However, in one of life's cute quirks, I awoke one morning to find myself in the belly of the beast, a radical/progressive teacher caught in the tightening coils of an ultraconservative institution, the public school system. Unwilling to take a public stance, I skated, slid, slithered and otherwise attempted to maneuver around the Pledge, unwilling to take a stance, but equally unwilling to expose my charges to this brainwashing, simply ignoring the directives of one administrator after another.

One day, however, perky little Ashley confided in me that her father "hates teachers who don't say the Pledge." Little Ashley's dad I knew to be an extremely large truck driver.

"Really?" I stalled for time. "Well, I'll talk to him," wondering what the hell I would say, should he be determined to expose me to the House on UnAmerican

Activities Committee. As soon as I had released the kids, I sat down at the computer, thinking, not for the first time, "Boy am I glad I'm a teacher of creative writing."

And I crafted a response, one that has withstood the slings and arrows of time. I said, in effect, "Dear Mr. Stossel: I am so glad that little Ashley has shared with me your completely understandable concerns about the Pledge. Let me assure you that no one holds this mighty nation, nor its sacred vows, in higher esteem than do I. In fact, in such high esteem do I hold this vow, this Pledge, that I feel obligated to do nothing that would undermine or trivialize its sacred nature. I would no more have children mindlessly mistreat and manhandle the flag to which we pledge, than I would have children mindlessly mutter a covenant they do not understand."

I paused for a psychic swig of brandy from my flask of courage. "In fact, I so value the Pledge that I devote quite a bit of curricular attention to internalizing and deeply understanding the nature and the gravity of this commitment. Many years of experience have taught me how imperative it is to approach significant concepts with steadfast planfulness and awareness, lest this sacred document and promise fall into the familiar dustbin of rote and meaningless recitation. We may certainly disagree, but I hope you honor my thoughtful approach to this important issue."

I received an appreciative response from Mr. Stossel, and breathed a deep and 'steadfastly planful' sigh of relief.

And to think, I could have been a politician.

Chapter 34
America's Army and
The Next Children's Crusade

As George Santayana warned, those who do not learn from the past are condemned to repeat it. In the Europe of Anno Domini 1212, in the midst of a dark, repressive time of fractured Empires and great social unrest (sound familiar?), children rose up to take the lead. Birthed and suckled by those horrid times, a children's movement erupted to rid the Christian world of the Infidel. Tens of thousands of children, mostly from France and Germany, whipped into a righteous war frenzy, coalesced for an assault on Islam to regain the Holy Land for the truly holy.

This "Children's Crusade", as it came to be known, seems to have initiated with a peasant boy of about twelve, Stephen. He had been worked up to a feverish pitch by the returning Crusaders (the priorday Veterans of Foreign Wars) and their processions and litanies for the retaking of Jerusalem. This epidemic spread throughout the pre-schools of Christendom, infecting kids as young as six years old. These little guys claimed to be prophets sent by

Stephen (no, this is not history according to Monty Python, this is the real skinny).

When the "prophets" had gained sufficient numbers, they began to march through the countryside, gathering strength in villages and towns. It spared no one, boys and girls drawn in almost equal numbers. Although many of them dropped out of sight during their march to Bethlehem, one thing is certain: of the tens of thousands of children who left, a scant few returned.

Which brings us to the Pentagon's new website, targeted directly at kids, called, "America's Army." The site offers a free video game described by the Army as: "A thrilling first-person game. Become a member of the world's premier land force: trained and equipped to achieve decisive victory anywhere. Earn the right to call yourself a Soldier, letting the enemies of freedom know that America's Army has arrived..."

The ad campaign calls it: "The official US army game. Live the real army experience, from basic training to special ops missions...No other army game is this real, because nobody gets the army like the army!"

"We want to get kids playing Army again," says game designer Dr. Michael Capps. "Kids play Nintendo, kids are online playing *Counter-Strike,* and we wanted to get in that space." There's a rifle range, an obstacle course, etc, and then kids are ready to join a squad, but only if they're properly trained.

"If you want to be a sniper, well, you've got to go to sniper school first. There will be slots available for snipers, but only if there is a trained sniper on the team." (Boy, that's a relief, isn't it?) Idea for a new ad campaign: Be all you can be. Be a sniper. Or what about a Sniper Magnet School? Sniper Charter School? Could be a moneymaker.

When asked about the topic of friendly fire (a hot topic

for "team-based" shooters), Capps said they're taking that very seriously. "Players shooting their own teammates will be kicked from servers, and repeat offenders run the risk of being banned completely (*bad* shooters!). We'll make sure we can provide a safe experience."

(I wonder if they'll offer free downloads of Michael Moore's *Bowling for Columbine?*)

Military and games reviewers are ecstatic about the potential of this free online recruiting tool. It's a highly competitive market and, as an advertisement says, "Second place is *dead* last!" A source no less expert than the German military says:

Ultrarealistischer Taktik-Shooter gepaart mit Propaganda-Mief: Wie spielt es sich als Soldat der amerikanischen Armee?

Although Jack Thompson, father of a ten year old and attorney representing parents of children killed in school shootings, is attempting to file an injunction in federal court, still the Pentagon is bullish about America's Army. After all, over one million people, many of them children, have already registered for the game.

Maybe soon our kids will be able to register online for the next Children's Crusade.

War Toys, anyone?

Chapter 35
"The Pleasure Palace"

M r. K, Mr. K," Gloria was framed in the doorway, flapping empty hands about, a seal out of water, "I can't get it".

"It's okay, honey, I rarely get it myself," I commiserated, dimly attempting to recall what 'it' was. Kids come and go regularly, irregularly, rather, from my classroom with high frequency. Some I've sent on missions, some have sent themselves, some may be on missions from other sources (these include the ever-popular 'George told me to go find his ******', to 'I was going to help Georgette *******', to mysterious instructions from what I refer to as The Land of the Possessed, 'I just had to *****'). The majority, I believe, are simply Bedouins adrift in the bleak deserts of public education, vainly seeking a refreshing oasis, or simply isolation.

Also, as sweet little demon faces morph into their parents, and hours bleed into days into years into bloody pools of decades, oozing around my stained and swollen feet, I no longer have neither the ability nor the interest to determine where others are going or why.

"You know," still waving about her flippers, "the tissue?" Gloria brought me back. That's right, I had sent her to the Bus Barn, the cavernous stronghold, the Keep, of the custodian slash bus drivers, on the trail of tissue to stem the free-flowing sinuses of the winter flu season. Why, a reasonable person might reasonably reason, are the boxes of tissue not readily accessible in a setting where flu vectors (otherwise known as sick children) are rarely kept home? Why, indeed. This is but a minor mystery in the mysterious labyrinth of public schooling.

"Weren't they there?" I shout across the class. I often engage in vain, rather stentorian attempts at communication over the heads, I mean literally over the heads, of my students, some 25 feet from overhead projector to doorway, always open, often occupied by someone such as little Gloria.

"No, they, I mean someone, was there but they were in the back, you know, behind the buses, and making weird noises."

Oh no, not again. We have been blessed/cursed with a parade of characters occupying the powerful throne of Custodian, sometimes consolidating power by also holding down the coveted 'Bus Driver' (or Head of Transportation) seat. As had happened with surprising frequency, a Bus Barn Romance had erupted again, fueled, perhaps, by intoxicating gas fumes and lubricated by 50 weight. They were humping away, once again, in the back room of the Bus Barn.

I'm not sure when the Bus Barn first became known as "The Pleasure Palace", or, for those of a less romantic vein, "The Quickie Lube", but it achieved a special prominence during the days of "Gorilla Woman". G.W. was a classroom aide who loved to play femme fatale. In a world populated by women wearing sensible shoes and paint resistant smocks, Gorilla Woman came to work dressed to the nines: satin spiked heels, stiletto, five inch, nylons, short flouncy skirts of gauzy organdy or chiffon, scarlet lipstick, and rouge. The G Woman cut quite a figure in the generally drab sartorial world of teaching.

Burly Burt, our bus driver slash custodian at the time, found he was not at all immune to G.W.'s feminine wiles, and the two would often be sequestered in the Pleasure Palace looking for "the right tool for the job".

The present denizens of the Quickie Lube were Dragon Lady, a short tempered, self-important, much divorced bus driver, here known as Dragonia, and a blockish, boorish, lout of a man, here known as Thuggo.

There was much speculation as to what Dragonia and Thuggo saw in each other, what they had in common. But this much was certain: prolonged exposure to the Pleasure Palace, it's oleaginous sheen iridescent in the rare moments of light penetration, the hard-edged raw masculinity of the highly organized Snap-On tools, well-hung from the walls, the perfumery of mingling scents: WD-40, diesel fuel, and lithium grease, all created a heady and dizzying sensuality, seemingly irresistible.

This much was also certain: they had the Bus Barn in common. You could almost envision them in a misty scene from Casablanca: as the school bus is about to pull away from the curb, a reluctant Dragonia leans out the window, lamenting to a crestfallen Thuggo, "At least we'll always have the Bus Barn."

"Mr. K, Mr. K", Gloria snapped me back, "What about, you know?"

I emitted a soft sigh as I reemerged from the unseemly image of sad, thickened souls foraging for solace in the gracelessness of the Pleasure Palace.

"I'll bring some Kleenex from home, Honey."

Chapter 36
Kindergarten, The Garden of Children
Where Have All the Flowers Gone?

As we educate our children to face an increasingly hostile world, we can happily and concomitantly, indoctrinate them and also make a profit. In addition to using the No Child Left Behind Act as a front to gain recruiting access to the schools, there is also the burgeoning arena of war toys, one of our most secure growth industries. Especially in times of insecurity, drugs and war always gain in popularity.

In the Pentagon's "America's Army" ("No one gets the Army like the Army"), you may recall, our children can go through Basic training, return (or initiate) hostile fire, be felled by "friendly fire", or even enroll in Sniper training. Dr. Capps, the creator of this user-friendly kid-recruiting video game, explains further. "You shoot a guy on the enemy force. He's got an AK47. He falls to the ground and drops the AK, but his friends see him drop an M16 because that's the good-guy weapon. You pick up his weapon and now you've got an AK weapon, with different capabilities than your own U.S. weapons. It makes for some interesting gameplay, but it's really clear when you throw people in." (I'm sure glad of that).

Ever since the Gulf War, there has been a seismic wave of domestic violence on U.S. military bases. The rate of reported domestic violence is five times that of the civilian population. First to capitalize on that trend is "Domestic Violence", a new game promoted by the Army because, "No one does domestic violence like the Army."

Still in the New Products Development Division of the military are games such as "Gay Bashing"(don't ask, don't tell), "Rape, Pillage, and Burn" (with a special command option to exempt you from the War Crimes Tribunal), a special issue "Vietnam Revisited" (your kids can actually destroy villages in order to save them; this game also comes with a hot, new Napalm simulation), "Cross Burning" (with a patented anti-Islamic option), and other games sure to stir your children's patriotic juices.

David Grossman, a former US military psychologist, explains in the *Toronto Globe and Mail* the interests of the armed forces in developing such games. Programs have been developed to "make new recruits more effective killers, to increase the trigger-pull ratio." The trick, the Army psychologist says, is to "break down the natural human aversion to killing." This is referred to as "disengagement." "The ability to watch a human being's head explode and to do it again and again--that takes a kind of desensitization to human suffering that has to be learned." And what better place to learn than in the safety and comfort of your own home?

Presently on the market are an armoried cornucopia of toys of mass destruction. JC Penney's Christmas Catalog features a takeover of Barbie's Dream House by GI Joe and friends.

There are also many anti-Muslim games such as the "Arabian Random Insult Generator" (as Dave Barry says, "I swear I am not making this up"). An online games reviewer gloats, "Amazingly, this generator is stocked with so many vocabulary words and phrases, there are presently 37,988,794,601,683,200 combinations of insults possible." Also offered "Choose Your Weapon: Shave, cut, shoot or nuke Osama bin Laden, and Bend Over Bin Laden: Shoot missiles at Osama", both by Newgrounds, "Beat the Crap Out of Bin Laden" from MadBlast, "Osama Hunt aka Kill Osama!" target practice from PoopyJoe, "Run, Taliban, Run" a shooting gallery from the Therapy Pages, and "Strip Bin Laden, find out what's under Osama's garb", also from MadBlast.

A net search revealed 650,000 links for "War Toys".

Among groups seeking peaceful alternatives are:
Physicians for Global Survival *www.pgs.org*
War Resisters League *www.warresisters.org*
Christian Peacemakers *cpt@igc.org*
Institute on Media and the Family *www.mediafamily.org*
Play for Life *www.usersglobalnet.com*
Toys for Peace *www.lionlamb.org*

There are choices. If we don't make them, others will make them for us.

Chapter 37

Pi Day

J ohnny, Johnny", I burst into MiniMac's office, "Let's roll".

John fixed me with *the stare*, vacant and aggressive both, that had helped make him an All-American wrestler. "What the hell are you talking about?"

"Let's get on the mat, buddy, you and me. Let's take it to the next level."

"I repeat," John tried to stare me down, "What the hell are you talking about?"

"You and me, baby, let's rumble. It's π day, kiddo. We're going to the mat, you and me."

"π day, oh shit I forgot. I didn't have time for that crap."

I was major, I mean, major, let down. "You mean you didn't practice?" I asked, totally sideswiped, completely undercut. "Do I understand you to mean that we're not competing? I am shocked, John, I mean thoroughly shocked. Shit, man, I've been practicing for a month. Are you putting me on?"

By John's bland, inscrutable poker face, I took it to mean that I had just spent a month practicing for a competition that was not going to take place. I was deflated, de facto deflated, flat. I limped out of MiniMac's office, looking for someone to take on, someone to compete against.

Once again, I had been set up. I had spent a month memorizing as many digits of π as I could, in the vain hope that I would have an opportunity to display my rather arcane knowledge of π. No deal.

John, the school math teacher, nay, math guru, had turned me on to an international competition/observance of π (the ratio of circumference to diameter of a circle) which occurred, reasonably and cutely enough, on March 14th, at 1:59 in the afternoon, in honor of π, the infinite (nonrepeating) decimal which begins 3.14159, or, 3/14 at 1:59 in the PM. In a blessing of synchronicity, the day is also Albert Einstein's birthday. How acute is that?

John had told me he was activating his rather formidable 'Math Club' to oversee and judge this event. Kids from Kindergarten on were incited to memorize as many digits of π as they could, regardless of whether they had the remotest inkling as to what π was. Not wanting to compete against the kids, John and I had set up a 'friendly' staff competition. John and I had been the only takers, and, obviously, I had been taken.

I had already taken some rather serious shit from friends and colleagues alike, as I embarked on my endeavor to memorize a non-repeating, infinite decimal, meaning one that has no pattern, one that makes absolutely no sense. I began my project on a long drive my wife and I took to see her parents. I printed out the first 100 digits of π from pi quest on the internet (where else?). I made multiple copies and had my wife hold one and feed me data as I drove.

Having memorized the first 64 digits, I called my friend Tom, the best math teacher I know, upon my return home. "Tom, Tom", I gushed, "You'll never guess what I'm doing."

"Building a patio," Tom guessed.

"No."

"Cutting slips of native plants to start."

"No."

"I give up."

"I'm memorizing π."

Silence.

"I've just memorized 64 digits."

Silence.

A longer silence.

"Why?"

"For π day. To compete in the competition. You know."

Tom measured his response. "I think if I had that kind of time, I'd memorize plant families, or bird families, or something useful."

If I had been deflated by MiniMac, Tom's response flattened me like an express train.

Nothing, if not stubborn, I tried to reach the part of me that had decided to pursue this meaningless activity in the first place. I realized several things. One, I was aging, and this was a personal quest to assess how aging had affected me. I had been a very bright kid, one capable of marvelous feats of memory. Did I still have it? Could I do it? I was painfully aware that time had indeed dimmed my skills, but, I wondered, what was left?

Two, I knew that memory games and activities played a very important role in staving off the onset of Alzheimer's disease. It seemed clear that those Seniors who played mind games, stayed saner longer. Worth the chance.

And Three, I am a firm Constructivist. This means that

I really believe that the only significant way we learn is to construct our own meaning. We simply don't learn, in any important way, information that is preformatted for us. We must personally format, reconfigure, information so that we can personally process it and make it meaningful to ourselves.

So, here I am, alone in this idiotic pursuit, at 314 digits (Get it? 314. Pretty cute). Yes, I know, that there are idiot savants who know 10,000 digits of π. Frankly, I don't give a shit. I'm doing the best that I can.

Chapter 38
Out, Damned Spot!
Boys will be Boys

Noah Shachtman reports in <u>Wired</u> magazine that, "The Marines' arsenal of the future is starting to look a whole lot like the shelves at Toys 'R' Us." The Marines have developed an auto-piloted airplane that closely resembles a remote control model airplane. The designer, Major Greg Heines, says, "If a Marine can use (Microsoft) Word, he can get this plane to fly." If that Marine can play Grand Theft Auto 3, he shouldn't have a problem handling the Dragon Runner, equipped with a camera and meant to be a mechanical scout in close combat situations. The runner isn't autonomous, however. A Marine remote controls the flat, 15.5-inch device with a six-button keypad.

"We modeled the controller after the PlayStation2, because that's what these 18-, 19-year-old Marines have been playing with pretty much all of their lives," said Maj. Heines. "It really saves us an awful lot of work. The fact that these kids are not only highly skilled by the time we see them, but also so incredibly gung-ho, we find really exciting."

Act V, Scene i

LADY MACBETH.

Out, damned spot! out, I say!-- One; two; why, then 'tis time to do't ;--Hell is murky!--Fie, my lord, fie! a soldier,
and afeard? What need we fear who knows it, when none can call
our power to account?--Yet who would have thought the old man to have had so much blood in him?

Col. Dave Grossman, a military historian and author of *On Killing*, argues that "the inflicting of pain and suffering has become a source of entertainment and vicarious pleasure rather than revulsion. We are learning to kill, and we are learning to like it."

Christian Peacemakers reports, "In fact, 25,000 Americans are murdered each year -- more Americans than were killed at the height of the Vietnam War. This increased violence is seen in schools, playgrounds and even churchyards. Children have a hard time distinguishing between fantasy and reality; in addition they receive very mixed messages from a society that glorifies violence through toys, movies and other entertainment -and at the same time tells them that violence is bad."

LADY MACBETH.
--What, will these hands ne'er be clean?
Will all great Neptune's ocean wash this blood from my hand?

No, this my hand will rather the multitudinous seas incarnidine
making the green one red.~

"We're finally getting over that Vietnam guilt trip", says Capps. "These kids, thanks to such early exposure at home and in arcades, have no hangups about wasting "bad guys". They not only see it as a blast, but also as kind of their patriotic duty."

LADY MACBETH.
Here's the smell of the blood still: all the perfumes
of Arabia will not sweeten this little hand. Oh, oh, oh!

No, we cannot ever wash the blood from off our hands. These are our children, reared by us and educated by us. From Howard Unruh, driven mad by World War II, driven to slaughter 13 of his neighbors in New Jersey, the first modern serial-killer, "I'd have killed a thousand if I'd had enough bullets". To Lieutenant William Calley, executioner of 300 women and children in My Lai, Vietnam, driven mad by war, exonerated by the President, hero of many. To the Columbine teens, trained by video games and fed up, unwilling to submit to ridicule anymore. To John Muhammad, trained by the military as a sniper, killing complete strangers, doing what he did best.

Our collective sense of outrage is natural, but disingenuous. These are our children, our creations. We must take responsibility for them, for ourselves, and begin to set some new examples, some new standards.

Children are not born violent--research shows that violence is a learned behavior. And it is a behavior that we are teaching to younger and younger children. The problem is not that our children aren't learning their lessons; the problem is they are learning them all too well.

MacBeth, Act V
DOCTOR:
This disease is beyond my practice.
Foul whisperings are abroad: unnatural deeds
Do breed unnatural troubles:
God, God, forgive us all!-

Chapter 39
The Roar of the Greasepaint

L adies and Gentlemen, is seeing believing? Can water flow uphill? You are simply not going to believe your eyes."

What had been billed as "The World's, or at least the Third Grade's, Greatest Magic Show" had proceeded as well as could have been expected. Jesse attired in a fluffy white shirt with red bow tie, looked a little more like a movie usher than a famous magician, but he was pulling it off. It's difficult to hold third graders' attention, especially while performing with small objects difficult to see in the best of circumstances. They will, of course, let you know, "I can't see, I can't see," even in anticipation, before there is anything at all to see. It's often a case of the blind leading the planning-on-being-blind.

Often the situation can be rectified by suggesting to the blinded child, that if he or she were to move his or her head two inches to the left, he or she would have an unobstructed view of the proceedings. Why it is necessary for outside intervention is one of my life's works little mysteries.

The show had run through a number of card tricks ("the four aces are hiding in the attic, waiting to rob the house"), several coin tricks ("this nickel has a mind of its own"), the infinite milk pitcher ("boy, I wonder if there's a drop left?"), and the dancing scarf ("that hankie really has rhythm"). It had seemed like Jesse had peaked too early when he sawed Lauren in half, an act replete with moans and curdled screams (I had axed the blood. Not appropriate, I ruled). Jesse was a better judge of his audience than I was.

He was really working himself into overdrive, like a used car salesman on high octane. "You are about to view the unviewable, the totally whacked out unbelievableness of," he paused to reel in the rubes, trying to drop his squeaky prepubescent voice an octave or so, "The Box of Doooooom!"

At this point, the highly theatric Jesse stepped aside as Lauren and Noah, all this while dancing like dervishes around the center of the stage, swept off a small tablecloth (actually a bath towel) revealing a fairly large cardboard box, about a three foot cube. The box was painted black, rather incompletely, Scott Tissue still showing through here and there, with rough letters of gold glitter spelling out, 'Box of Doom'.

"Ooh, ooh," chanted Lauren and Noah, trying to whip up the crowd, alternately striking poses with an open hand stretched out at the box, then a beckoning hand stuck out at the audience, then a few dervish spins, as they traded places on either side of the box. "Ooh, ooh."

"Behold the unbeholdable as my assistants and I plunge the Swords of Doom into the Box of Doom, defying death and," Jesse paused for a moment, glancing at Charlie and Adrian for help, fumbling then recovering, "and various other things."

With great ceremony and pageantry, Jesse handed out

foil covered yardsticks, the Swords of Doom, to Charlie and Adrian. As the dancing duo of Lauren and Noah fell back to a safe distance, adopting Vanna White poses, the three fell upon the box with fervor, grunting out 'Gah's with each stroke as they thrust their swords into the Box, slowly penetrating as if the cube were made of some extremely dense material. Occasionally a sword was withdrawn as the assailants sought another slot in which to plunge. Leaving their swords protruding from the box, they would pick up another, seeking out a different angle, and plunge again.

The audience began to wiggle and fidget, some looking for soft spots to poke, until they were brought back to rigid attention by what seemed to be moans emanating from the Box of Doom.

"Unh, unh, oh, unh."

Almost inaudible at first, they rose and fell with an eerie rhythm.

"Unh, unh, oh, unh."

The dancers, Noah and Lauren, looked appropriately frightened. Jesse, Charlie, and Adrian eyed each other, ceremoniously picked up their last swords, and, with synchronized precision, drove them deep into the box. The moans rose in volume, hovered, then ceased. Silence hung over the class.

With face drawn, tight-lipped, I strode slowly over to the Box of Doom, withdrew the ten or so swords, and opened the top. Looking in I saw Zach, miniscule Zach, huddling in the one safe corner of the box. I allowed relief to flood my face, as Zach uncurled his limber, little self and stood up, bowing to a tremendous ovation.

Smiles all around, as I congratulated the magicians and performers, the audience for its behavior, and released the class to recess.

155

Frail, petite Dominique, who had not actively participated, but seemed to have been engaged, came up to me, meekly as always. She gave a light tug at my sleeve until she caught my eye. "Mr. K., third grade is really magic."

I held her eye with a slight nod, "Sometimes it is, Dominique, sometimes it is."

Chapter 40
The Military Takes to the Schools Like a Leach to Blood

Given the choice between mindless, knee jerk complicity with the military, and conscientious objection (a thoughtful opposition to war and violence) give me the latter. Given the not-so-hidden coercion of the No Child Left Behind Act to procure the access of military recruiters to school campuses and personal records of the students, as well as the rather insane drive to include warcraft as an academic field of study, we are confronted with the choice between military complicity and conscience.

Many school districts around the country are standing up to the federal government's drive to militarize our schools, from recruitment to JROTC, junior reserve officer training corps. Traditional peace churches have forged alliances with children's advocacy groups and community-based organizations to combat this out-of-wedlock merger of military and education. The military mindset is grudgingly difficult to change, but we need not be so complicit.

When did militarism become synonymous with patriotism? At what point did we, as a culture, decide that violence displayed love of country, and peacemaking was antipatriotic? It is not law, rather it is unthinkably misguided policy, an assumption that all too often goes unchallenged. Two hundred years ago, Daniel Webster took the floor of the Senate to challenge the militant God and Country link. He demanded, "Where is it written, in what section of the Constitution, that we may take parents from their children and children from their parents and compel them to fight in any war in which the folly and wickedness of the government may engage us?"

Thomas Jefferson replied in agreement, "Forcing people to serve in the country's army is the worst form of oppression." And, I might add, forcing the public schools of this country to mindlessly accede to the demands of our increasingly militant state demonstrates both "folly and wickedness." The consent of our schools demonstrates all of the above, as well as criminal cowardice. If the Founding Mothers and Fathers of our country were alive today, they'd be raising Hell.

These ROTC classes, these junior commando-sniper training classes we offer in our schools, what is it that they teach, as they drill their young uniformed cadets, our children? As part of their indoctrination do they expose them to the views of the Office of Naval Intelligence which said in 1947, "Realistically, all wars have been for economic reasons. To make them politically palatable, ideological issues have always been invoked. Any possible future war will undoubtedly conform to historical precedent."

Do they share the views of General Smedley Butler, Commandant of the United States Marine Corps, who said, "War is just a racket. Only a small inside group knows what it's about. It is conducted for the benefit of the very few at the expense of the masses. The trouble with America is that the flag follows the dollar and the soldiers follow the flag."

Do they invite General Butler, Boss Marine, to address their classes to share his experience? "I spent thirty-years and four months in active military service as a member of our country's most agile military force. I spent most of my time being a high class muscleman for Big Business, for Wall Street, and for the Bankers. I was a gangster for capitalism. I wouldn't go to war again as I have done to defend some lousy investment of the bankers."

Does The No Child Left Behind Act, which rails against unqualified teachers, address the JROTC classes, more often than not taught by uncredentialed teachers, the only such classes offered in high schools? As No Child Left Behind unabashedly pushes Bush's corporate pals' phonics programs, does it address the JROTC curriculum? When the Marines in Albany, NY could not provide a textbook for the class, they used the standard first-year Army JROTC textbook which claims, under the section on "brain power", that whites are said to be "left-brain individuals" who prefer "being on time" while "right dominant" African Americans prefer "a good time." Racist Aggression 101.

Eugene V. Debs said it most succinctly one hundred years ago, "I would no more teach children military training than teach them arson, robbery, or assassination."

Chapter 41
Bad Names

J ustin, Justin's a bummer," Ernie hunched forward in his blue plastic cafeteria seat, leaning eagerly on the staff lounge table, "Never, ever Justin."

"Jonathan--That's mine, Jonathan. John's okay, in fact, John's not bad at all, but Jonathan...phew," Ray almost spat in disgust, her nose wrinkling.

I surveyed the teachers sitting around the table, all having struck the same pose, reflectively searching the ceiling tile, deep in thought.

The game we were playing was: Names I Would Never Name My Kid and I Would Heartily Recommend That No One Else Use That Name Either, Because Every Kid I've Ever Had With That Name Has Been a Bummer.

"I just saw in the paper," Ray continued, "That Jonathan Marley's in jail. Stole a car, I think."

Everyone nodded sagely, surprised it had taken so long for Jonathan to 'graduate' to higher crime.

"Yeah, but remember Jonathan Sheldon? A dream kid, one of the best," protested Graham.

It wasn't enough to have had one difficult Shlomo, or one impossible Theresa. All Shlomos and all Theresas had to be tough in order to qualify for the Bad Name list.

"But when he was younger," Ray never let go. She was a bulldog. "When he was younger, he was called John." She also had a fabulous memory for detail. "He changed to Jonathan in fifth grade."

"Yeah, but he was born Jonathan." Graham was used to disagreement with Ray. Everyone was.

"People, people," I pleaded, whining imploringly, "Can't we all just get along?"

"No!" Ray barked flatly, "Not as long as Graham's full of shit."

"Listen, listen, everyone's right. It's completely subjective. One man's poison is another man's wife."

"Screw you, Guy, you sexist hog."

"No, really, you're all right, even Graham, Ray, but you're also all wrong," I paused for dramatic effect.

"Tim," I announced, giving time for the single syllable to sink like a piece of rockfish bait. "Tim...think about it." Brows knit in thought, straining to burst through the barrier of repressed memories, dragging the bottoms of their minds like a police net dredging the bay for a corpse.

One survives in teaching by repressing the worst memories and truly believing that hope springs eternal, but this Name that Nightmare game required that we retrieve, relive, even, the worst of times.

"A Tim graffitied the school, remember 'Lake View is Gay'? A Tim molested his little sisters and was barred by court order from his home. It was a Tim smoking pot in the Arboretum. And," I paused, glancing around the table, reeling in my catch of the day, "It was a Tim," my voice rising, "Who pissed all over the wall of my tent, *my tent,* on the camp trip. Too many Goddamn Tims." I pounded the table for end punctuation.

Graham looked over at Ray, almost warmly, and offered, "Wasn't it a Tim who lit the fire in the waste basket in the boy's room in your hallway?"

Ray replied, "I think, the kid who cut through the live extension cord in your room, wasn't that a Tim?"

Graham smiled, nodding in fond recollection, "I just heard a loud pop, then smelled the smoke. I never saw a thing."

I bathed them all in my embracing grin, pleased to bring everyone together. "Too Goddamn many Tims," I pronounced.

"Amen," all throats issued forth.

That night, as I poured my third glass of wine, I winced in anticipation. I had just seen my class list for the coming year. Two Tims in my class. I unsteadily raised the glass to my lips. It was going to be a long year.

Chapter 42
Just Say No

As the storm clouds of war and universal distress gather and darken, we must ask ourselves how we choose to respond. How often are Peace classes taught in our schools? How much time is spent on mediation, and non-violent conflict resolution? How much energy is expended by our administrations denying equal access to active peacemakers and counter-recruiters?

Do our schools believe that the ROTC instructors and the gung-ho military recruiters, pursuing their recruitment quotas as avidly as used casket salesmen, are really the most ethical, most well-informed models we want to provide our children? Perhaps the present-day Vets for Peace, or the General Butlers, or the Jeannette Rankins, or the Daniel Websters, or the Abigail Adamses, or the Thomas Jeffersons, or the Gandhis, or the M.L. Kings, or the Philip Berrigans, or the Amy Goodmans, or the millions of heroes who toil for a world of tolerance and non-violence, perhaps these folks are simply unqualified to teach our young.

Yes, these are indeed trying times. Yes, the hounds of war are loosed and packing, with a mind of their own. Yes, Death is at the door, but must we insist on making its job so damn easy?

Conscientious Objector
Edna St. Vincent Millay
I shall die, but that is all I shall do for Death.

I hear him leading his horse out of the stall;
I hear the clatter on the barn door.
He is in haste; he has business in Cuba,
business in the Balkans,
many calls to make this morning.
But I will not hold the bridle while he cinches the girth.
And he may mount by himself: I will not give him a leg up.

Though he flick my shoulders with his whip, I will not tell him which way the fox ran.
With his hoof on my breast, I will not tell him where the black boy hides in the swamp.
I shall die, but that is all I shall do for Death; I am not on his payroll.

I will not tell him the whereabouts of my friends nor of my enemies either.
Though he promise me much,
I will not map him the route to any man's door.
Am I a spy in the land of the living that I should deliver men to Death?
Brother, the password and the plans of our city are safe with me; never through me
Shall you be overcome.

****

As we enter into a new millennium locked in a battle to the death over when best to introduce the semi-colon and long division to increase test scores, the real battle for our children's souls rages about us largely unheeded and unattended. While a wolverine runs wild through the nursery, we debate the pattern of the children's wallpaper.

Just as surely as there are community-skill classes and vocational-skill classes, there need to be personal-skill classes. Classes that give young men and women the opportunity to discover who they are, what they hold dear. A school that does not offer classes in peacemaking, conscientious objection, non-violent conflict resolution, pacifism, the culture of militarism, the histories of inequities in race, religion, gender, and gender preference does its clientele and the community no great service.

Yet we persist in inviting the military into our schools. Inviting, promoting, glorifying, and even subsidizing. To what end?

As General Butler observed, "Like all members of the military profession...my mental facilities remained in suspended animation while I obeyed the orders of higher ups. I never had an original thought until I left the service."

For "service", substitute "school."

Chapter 43
Noah

O
h, no, not Noah!"
Images of a cherubic face, broad of mouth, frog-like, even, but absolutely adorable.

The first time I saw Noah, he was in Kindergarten, skipping (Noah never just walked), skipping to the lunchroom. As he passed my open door he stopped, flashed his beatific frog smile at me, and blew me a kiss. Holding eye contact, he pressed two fingers against his lips, held his little hand palm up in front of his face, and blew. It was almost tangible, this little guy with his little, big heart, blowing a tiny winged love right at me.

A five year old, boy at that, blowing you a kiss, is not something that happens every day. What a guy!

Three years later, I was lucky enough (okay, it was parent request) to have little Noah placed in my third grade class. Gifted with an immensely vivid imagination, he was a lazy puppy. Charming and mischievous, but lazy. After school, Noah would hang around, bat his long, blond lashes at me, and I'd respond, "Okay, little buddy, you dictate, I'll type."

And I'd sit at the computer, open up one of his documents, read aloud what he had already written, and say, "Okay, shoot."

Those key words, that shibboleth, would trip Noah into a trance, some kind of artistic dream state. He'd lean back in his chair, scrunch his eyes together, tilt his head back and relax. Not like he wasn't relaxed before, Noah was always rather kicked back, but at times like this, he melted. A small smile spread across his beaming little face, like a shaft of morning sun spreading across a stretch of sand, and he'd start to talk.

"Back then, in those long ago days, before there were pencils or paper or guys, there were only ladies. Ladies were the only people."

"Wait, a minute. Okay, go on."

"Well, in those days, these ladies and the snakes," Noah lisped disarmingly.

"Snakes?"

"Oh, yeah, the only other things were snakes. But in those days, oh, so long ago," Noah opened his eyes, winked at me, and continued, "Oh, best beloved, the Earth was a cube and the poor ladies..."

"And the snakes?"

"Yeah, the ladies and the snakes couldn't sleep."

"Why, not?" I drew him out, unnecessarily.

"They couldn't sleep because of all the banging and clanging," he glanced slyly at me to see if I noticed his rhyme scheme, "the bashing and clashing, the smashing and snashing," and Noah was off, way off in his own lovely world. I simply served as his scribe.

****

"Oh, no, not Noah!"

"Yeah," Andrea was spreading the news, "I don't know all the details, but it seems like he was trying to cross Central and never saw the van."

"Where is he now?"

"He's in a coma. They flew him down to the City. That's all I know."

My heart turned to lead, and dropped right out of my body.

The day before school, walking home from festive back-to-school shopping, Noah had been hit by a pick-up, suffering serious, perhaps irreversible, brain damage. It was harder than usual to begin school that year. He should have been entering Junior High, instead he was comatose in Oakland Childrens' Hospital, 300 miles from parents and friends.

****

Four months, four long months Noah lay in a bed in Oakland in a coma. His parents, emotionally and physically drained from their seven hour commute (Dad and Mom would trade roles, literally pass on the freeway about 150 miles north of the city, like Emperor Penguins passing a fragile egg, balanced precipitously on clenched toes, from male to female), his parents pulled him from the hospital, still comatose, and brought him home. Linda, his ferocious mother bear of a mother, refused to admit defeat. She called me up asking that I take him on as a student, every day after my normal workday.

I couldn't say no, although I hadn't the vaguest notion what to do with a kid in a coma. I quickly read up on brain trauma, but was thoroughly ill-equipped to deal with the reality.

I walked into a house, pretty much in total disarray

from five months of neglect and distraction. It took a moment for my eyes to accustom to the dim light. Finally I saw him, slumped over in a wheel chair, listing to the side, his tongue lolled out of the side as a ball of spittle formed at the corner of his mouth, dripping rhythmically onto the arm of his chair.

"Noah, honey, Guy's here. Remember Guy, your favorite teacher?"

His head jerked, rotating crookedly up. He didn't meet my eyes. Did he see anything at all? I couldn't tell. I could feel my face freeze into a rictus of friendliness. My hands felt clammy as I tightly gripped onto the few, pitiful teaching supplies I had brought. My tongue was stuck to the roof of my mouth as I swiveled my head frantically around looking for a retreat, but his mom stood behind me, blocking my exit. I was terrified.

Chapter 44

Why Teach?
Who Are These Fools?

My good buddy, Jim Steinberg, began his life as a lawyer. No, he wasn't born that way. What a frightening thought.

Mom and Dad Steinberg: "Well, Doctor, is it a boy or a girl?"

Doctor: "I'm sorry, I don't know how to tell you this, but...it's a lawyer!"

Mom and Dad: "But how could this happen? We took the utmost precautions."

Doctor: "Sometimes even the finest prenatal care can't prevent these genetic sports. We, in the profession, feel that sometimes Nature toys with us."

No, little Jim began his *professional* life as a lawyer. When his reason started to slip away, he became a teacher. Drawing upon his legal background, especially as a defense lawyer, Jim used to refer to Back-to-School night as 'The Opening Argument for the Defense.' And it often is.

Teachers are commonly on the defensive in our culture. Now, how did this come to be?

Yes, teaching is stressful. By independent audit, teachers make over 1200 decisions at work each and every day. This resolves to around 200 decisions an hour, or over 3 per minute. Well, you've heard it here first: they're not all right, and they're not all popular. And it's not just the quantity of decisions, it's the quality as assessed by others. Our decisions can have enormous impact on the kids around us, enormous and, too often, unforeseen.

It was my high school art teacher (being prep tracked I was only allowed to take 1 art class in my 4 years of high school) who so demeaned my talent that I refused to draw for the next twenty years. It was my 8th grade teacher who, mistaking me for the atonal child seated behind me, ordered me to 'mouth the words' during chorus. Guess, what? That was the last music class I ever took.

Everyone has school horror stories, some so painful they cannot be shared or even accessed. Repressed school molestation memories, acts of psychic violation so dreadful they can only be excavated with the aid of trained professionals. "Guy, Guy, I want to talk to Guyzie, the injured artist. Guyzie, are you in there? Come on out. It's safe to talk, now."

Conversely, other molestation incidents are so painful they must be shared. These are educational assaults so intense we must share them to exorcise the demonic damage. "Grandpa, tell us again about the time the whole class broke out in derisive laughter when you mispronounced that word during class read-around time." For me it was sepulcher, a word I had seen but never heard pronounced. Sepulcher. Forty years later I remember the word, but have forgotten the teacher. But the laughter still rings in my head.

For others the elemental remembrance of simply going to the bathroom brings on the sweats. Our potties were so primitive, so primordial, that were used as the set for the film, "My Bodyguard", filmed on site at Lake View HIgh School in Chicago. Constipation was common.

All of us cart around our schooling baggage, old beaten bags bearing the scars and lacerations of our traumatic journey through the school system. Bags that bear constant reminder of how vulnerable and fragile we all are. And it's these reminders that inform and caution us about our behavior with kids. Our interactions are fraught with the fear of inflicting unwitting damage, scarring our charges as we have been scarred.

In fact, at some point late in the year, I corral strangers: at gas pumps, in the supermarket, at the vet, at the marsh, and always ask them the same question, "Is your job stressful?" Some have recoiled, but most, seeing me as a threatless middle aged male, have responded in the affirmative. Teachers should not try to corner the market on stress, however, assessments over the years have placed teachers in the top 5 in vocational stress, consorting with cops, firemen, and air traffic controllers, jockeying for top stress post.

In the good old days, politicians used to rank in the upper echelon, but in recent years they have been too stupid to be stressed by their positions. (What, me worry? Bombs away).

The stress derives from the basic fact that the job description is impossible, unapproachable, unreachable. Meet the needs of 20, 40, 150 beings, incredibly diverse, all with immediate, intense, and personal needs which must be met, and met now.

Embedded in this job description is built-in failure. The job is impossible to do perfectly, or even well enough. All teachers, even award-winning educators, even those with thirty, forty years of experience; all teachers worry, fret, and bedevil themselves with the painful realization that we will fail. We cannot do our jobs as well as we wish to do them.

Stressful? Yes. Could we do better? You bet. Would we choose to do this again? In a heartbeat.

Chapter 45
Noah 2

My first encounter with the head lolling, slack mouthed, totally defocused Noah, left me wanting to escape his house as quickly and cleanly as I dared. How in the name of God, I wondered, could I possibly face Noah again? How could I face his parents? How could I face myself, knowing I could not possibly do him any good? I, charlatan that I was, slunk off.

I had, however, promised Linda, Mama Bear Extraordinaire, that I would be there for him, for them. It was time for a little chat with myself.

"Guy"

"Yes"

"Listen. Tom and Linda didn't choose you to work with Noah because of your vast store of experience and success working with head trauma clients."

"I know."

"You know why they chose you?"

Hanging my head, shyly scraping my shoe on the soft, red clay.

"Yes, I know."

"So, go ahead, tell me."

"Well," I wheedled, "They chose me, not because of my expertise, but because of my love for Noah, actually, our mutual love. Also, they probably thought that his familiarity with me might rekindle his memory. Maybe jumpstart programs that have been stalled."

"So, okay," I agreed with myself, "That sounds like reason enough to do the job. If you don't know the dance steps, just put on the music and feel the beat. No harm, no foul."

Realizing the severity of the case, that I couldn't possibly do any further damage as long as my intent was clear, I talked myself into continuing the work. I really had no option, anyway.

Day in, day out, I'd shut down my classroom by 3:30, and run up to Noah's house for our 4 to 5 o'clock session. At first, I'd just sit opposite him, talking about the good old days, cooing gently, or showing some excitement, always well modulated, well tempered. Noah tended toward agitation. Although generally he did not really appear to 'be there', he was certainly short tempered. Linda had asked me to bring my guitar and sing songs with him, try and access his longterm memory. Open some files. Encourage him to sing along.

Most of my attempts to engage him, however, met either with complete indifference or agitation. He did not speak, but he did growl, bark, and quiver when aroused. It was pretty unnerving. I tried to distance myself, not take his outbursts personally, but it was difficult. All of his caregivers agreed that anger was an important step in his rehab; at least it showed that things were getting to him. We assumed that his agitation derived from frustration at his fractured fragile body's inability to respond as he, the intact he, wished.

Then one day as I entered his dim sanctum, he tilted his head toward me, feebly raised his one good arm a few inches, and said, "NNha". My, God, he's talking!

Linda stood alongside his chair, beaming at me.

"Hi, Noah," I offered cheerily.

"Nnha, Gaa," he repeated, slowly, tortuously drawn out, the densely nasal sound issuing forth from a dark, serpentine tunnel, dank with fungus and lichen, deep and forbidding. "Nnha, Gaa", curling his lip in what I assumed was his best attempt at a smile, appearing to be pleased with himself.

"Hi, Guy," I repeated, idiotically. Then again, "Hi, Guy."

He cocked his head at me, virtually parallel to the ground, and peered out at me from one eye, an almost reptilian look. And, as is often the case with reptiles, he revealed a glint, a shard, of finer intelligence, a hidden intelligence, held captive deep within this almost grotesque outer form. For the first time in over half a year, I realized, rather than just hoped, that Noah was indeed home. There was a him, there. My eyes misted over, but I didn't want to look away, afraid that he would disappear. His mother, She Bear Extraordinaire, was right. If we joined forces, family, friends, loved ones, we could overrule the experts. We could will him back.

"Nnha, Gaa," said Noah once again, now evidently pleased with the result, perhaps thrilled that he had reduced his once loquacious teacher to stunned silence.

The hour upon hour, day upon day, month upon month, of seemingly useless activity, fruitless entreaties to sing songs with me, play board games with me, listen to my stories, look at me, for God sake, just look at me. All that time just melted off me in that one rush of realization: Noah's coming back! Thank God.

Chapter 46
Fred Would Be Pissed

I couldn't watch Mr. Rogers when I was young. He simply drove me nuts. His syrupy ingratiation into the world of children made me distrustful. His seemingly saccharine smile and welcoming warmth must be fake, I reasoned. I was really more comfortable with Eddie Murphy's hip, ghetto takeoff on Fred, who encouraged, as he leaned leeringly towards the camera, ("Hey, little boys and girls, can you say M_____ F_____?"). Or a line from my parents' early "Radio Bloopers" album, an aside from Sweet Stan, the Toddlers' Buddy, after he had finished the Sweet Story of the Day, "That ought to hold the little bastards 'til tomorrow.")

But when all was said and done, and I had aged and mellowed, not quite like fine wine, more like an old stinky cheese, Fred Rogers was right. He was good and true and constant in his acceptance of all kids. Just before Fred died, he recorded a spot for PBS. His soft and world-saddened eyes reached out at the camera in that Mr. Rogersian way that made each kid feel he was speaking just to them. Hell,

I thought he was speaking to me, too, when he said, "The least and the most we can do, is to let our children know that we will take care of them...no matter what."

Fred Rogers was a thoughtful, intelligent man. He did not say, "We must let the children of the Obscenely Wealthy White Old Men know they are cared for," nor did he say, "We must let American children know they are cared for," for sweet, old Fred Rogers had become a citizen of the world. Fred, in his accepting, childlike way, saw that for our planetary culture to survive, we must truly be willing to care for each other, American and Iraqi, Black and White, Muslim and Christian. Oh, my God, how we have failed.

As our campaign of economic sanctions creates an Iraqi public health nightmare sufficient to kill over a million Iraqi children in the last decade, as our "Smart Bombs" snake their way down a ventilation shaft and incinerate 500 sleeping women and children in Baghdad, as our depleted uranium devices continue to breed cancer for decades to come in Iraqi and Americans alike, as a U.S. gunship hovers over a wedding party in Afghanistan, slaughtering hundreds of celebrants, as we rain terror, unimaginable and soul-staggering, upon a huge city of children, what would Fred Rogers say? Would he say that we are acquitting our responsibility to "take care of our children...no matter what?" I think not.

How is it that our educational system has taken a decent, basically big-hearted people and warped them to such an extent they support the wholesale slaughter and assassination of children? How is it that our educational system has created a citizenry so ignorant that half of us believe that Saddam Hussein was responsible for the bombing of the World Trade Center, something not even claimed by the pathological liars of the present administration? How have we created a citizenry so ignorant that more than half believe that Saddam has supported and trained Al-Qaeda fighters, so totally ignorant of the traditional antipathy between Iraq's secularism and Bin Laden's fundamentalism?

How have we taken care of our children? We have taken the bright jewels of youth, the brilliant, multi-faceted gemstones of childhood, the exuberantly creative and questioning minds of the young, and systematically tortured the artistic opulence out of them, standardized, tested, and graded them into uniformity and compliance. The wild extravagance and critical divergence has been ground down to a dependable, dull, and dutiful populace that will sanction the wholesale slaughter of others.

We must create a new High School Exit Exam that does not release Americans into the workplace until they have demonstrated a critical thinking capacity to make thoughtful decisions on civic matters. We should hold our youth in preventive detention until they have demonstrated a commitment to social justice and basic decency, and may be safely released into the body politic. We've got to hold the educational system accountable, not for accurately teaching simile and Boyle's Law, but rather for creating an intelligent, independent, creative, and compassionate citizenry.

Oh, Fred, we are failing. We are not taking care of our children, neither ours, nor those of our planetmates. We're sorry.

Chapter 47
Noah 3
Kiss Every Frog

The origin of this allegory is shrouded in the swirling mists of meta-history, but I would like to ascribe it to Loren Eiseley. Dr. Eiseley, a brilliant paleontologist and one of the finest nonfiction stylists who ever picked up a plume, tells the story of the Star Thrower. Eiseley was strolling down a beach at dusk when, in the far distance, he saw a dim, shadowy figure stooping over, straightening, and stooping again. Loren continued down the beach, stepping over starfish after starfish that had been washed up. He picked his way over the starfish until he eventually encountered the man. As he approached he saw that the guy was slowly making his way down the beach, bending over, picking up a starfish, and flinging it back into the surf.

Eiseley greeted the guy and, although obvious, asked him what he was doing.

He replied, "I'm returning starfish to the sea."

Eiseley observed, "I just walked down half a mile of beach and must have passed tens of thousands of starfish.

How can what you're doing possibly make a difference?"

The man continued down the beach, stopped, picked up another starfish, and flung it back into the water. He looked at Eiseley and smiled, "It made a difference to that one."

Years ago, Andrea Suttell, my wonderful teaching assistant, and I were strolling down the sand at Agate Beach on a school camping trip, when we encountered a mass of jellyfish laying on the sand. Sail by the Wind, Vellela vellela, sometimes are caught by unfortunate tide and wind conditions and pile up on local beaches. I bent over, picked up several, threw them back into the water and related the story of The Star Thrower to Andrea.

Within a week she had painted me a marvelous watercolor of jellyfish, and presented it to me with the caption, "One jellyfish at a time."

How does a teacher deal with the overwhelming task of meeting so many kids' diverse and pressing needs? How does a teacher take on the job of overcoming years of dysfunction, years of inattention, possibly years of neglect or abuse? How does one do it? One kid at a time. One jellyfish at a time.

****

It has been 4 years now since Noah's near fatal accident. Over the tortuous and arduous weeks, months, and years I have seen Noah transform himself from a deeply comatose husk of his former self, into a new being, resolute, vital, intelligent, funny and happy. He has relearned the basic skills of swallowing, eating, speaking, walking, and relating to the world in an inquiring and bemused fashion, mischievous as ever, very much like his former self, but also different and newly mature.

His family was absolutely right on. The skeptics of the

modern scientific community can be overwhelmed by the power of love, accompanied by a massive dose of courage, integrity, and steadfastness. If you steer your own course with resolution and faith, you will be rewarded.

Teachers are often heard to say that they have learned more from the kids than the kids have learned from them. Often this is mere lip service, but in Noah's case it could not have been more true. I think of Noah and the starfish and wonder, who is the starfish, and who the thrower? Does it really matter?

Mahatma Gandhi used to say, "Nothing we do makes any difference, and yet you must live your life as if everything you do makes a difference."

Or as primary teachers are wont to say, "Kiss every frog. You never know when one might be a prince." I say, to hell with the princes. Just kiss every frog. They deserve it. And you just might find it changes you, deeply and forever.

Chapter 48
The Better Baby Institute, or No Toddler Left Behind

Finally, at long last, George Bush and a coven of his father's friends (what kind of guy has no friends of his own and is forced to hang out with friends of his father?) have launched his new initiative, the "Test the Babies Initiative". Otherwise known as the National Reporting on Child Outcomes System, this bold new approach to human assessment has the Federal Government providing a high stakes test to the 1 million preschool children enrolled in Head Start programs nationwide.

This attempt to "Go Where No Damned Fool Has Ever Gone Before" is truly an example of thinking outside the box. It also provides an example of thinking outside the body, a kind of out of body, out of mind experience. "This is necessary to ensuring (sic) that every child is progressing the way that they should," said Windy Hill, head of the Head Start Bureau that oversees the nation's largest early childhood program. Head Start provides social and educational services to low-income preschoolers and their families. "This would allow us to target our resources."

The new reporting system would require that all Head Start children be evaluated in the fall by a "battery of assessment instruments", and again in the spring to measure improvements. The proposal dovetails nicely with the high-stakes testing program already in place for grades 2-12. "Now we've got to fill the Kindergarten and First grade gaps in the system," said Ms. Hill.

Bush's Child Outcomes System is necessary, administrative sources feel, to provide standardized data to allow them to evaluate local Head Start programs, allowing the government to target resources to programs that were "underproducing." "The message we wish to send," says Ms. Hill, "Is that it's never too early to be held accountable. If you are receiving government funding, you must show results."

Head Start leaders and early childhood researchers are critical, feeling that a high-stakes test for toddlers will result in unnecessary stress and provide very little useful information. "Young children are poor test takers...and have a restricted ability to comprehend assessment cues," said Samuel Meisels, president of the Erikson Institute, a non-profit that trains child development professionals.

Hill responds, "These so-called intellectuals are pandering to the children and to the programs that take government handouts and simply play with the children. Face it, high-stakes testing is a reality of the modern landscape, and if you don't want to test the kids, we'll find people who will."

In addition to the Toddler Test Battery, the Bush administration has also placed on the table a final preschool exam, NPREE. The National Preschool Exit Exam has been proposed to ensure that preschoolers have the requisite skills to succeed in today's ever-more-competitive Kindergarten Rat Race or KRR. When asked what would happen to those who failed the Preschool Exit Exam, Hill responded, "Keep them in preschool. It would be irresponsible to unleash these ill-prepared children on an already saturated job market."

A high-ranking government source suggested that yet another test was still in the planning and development stage, modeled on the military vocational and aptitude test, tentatively titled, "The Early Childhood Aggression Index Assessment," or ECAIA. This test would measure the precocity of kids to become soldiers, allowing the government to pinpoint resources for tomorrow's armed forces. "This Warrior Assessment Rating System, or WARS, will allow us to separate the men from the wimps at a very early age. An extremely sophisticated weighted scoring system will not only rank the children on aggressive tendencies and response to actual combat situations, but also allow kids to earn points for contributing important security information about their families and play groups."

"Only the swift and the cunning will survive," observed Hill. "We see preschoolers as the new Mobile Assault Team. They need to be lean and mean, able to be deployed at an instant's notice, traveling fast and light, living off the land."

When asked if this would not only encourage violence, but also encourage kids to spy on their parents, an administration spokesperson responded, "Well, that's the point, isn't it?

Chapter 49
Into Thin Air

"Mr. K, Mr. K," a few girls interrupted my reveries. It was DEAR time, Drop Everything And Read time, and I had made myself comfortable in my soft swivel chair at the front of my class. Actually not the front, my classroom never had a front, it was more theater in the round. There were no desks to face forward, and no teacher's desk.

I had discovered early on that a teacher's desk was simply a barrier between teacher and student. It overly reminded me of our boss lady's marble slab desk top, more suitable as a sarcophagus, really, that separated boss from employee, superintendent from lowly teacher, the goddamned Almighty from the lowlife, a separation of powers. More specifically a separation of those with power from those without. So I had sound pedagogical basis for discarding my teacher desk.

On a less lofty ethical level, I also discovered that teacher desks are shit magnets. They attracted crap for which you couldn't find a good place. Stuff not easily sorted or filed. And it was the one spot in the classroom that could remain as it was until end-of-the-year cleaning. Dealing with the mounds of debris that accumulated would generally take several days after school adjourned for the summer. I would launch into a desktop archaeological dig in June, cursing my colleagues who were off to the beach, gin and tonics in hand, while I probed, dusted, discarded, or finally filed the year's worth of debris that my damn shit magnet had attracted.

Embarrassed at the unkempt state of affairs I had tried different techniques. I had followed the lead of Louise, a similarly desk-challenged colleague, and draped a sheet over my desk for school functions such as Open House and Back-to-School Night. A few parents had asked about the shrouded desk and I had responded that I had a top secret project sequestered under wraps. I had raised two of the legs off the ground so that any pile greater than 12 inches or so, simply slid off onto the floor, creating a strong disinclination not to pile. This had led to the loss of some papers, most notably a student's copy of the sacred CTBS test, the Comprehensive Test of Basic Skills used to track student progress. I had been severely reprimanded for that misdemeanor. In the end, I simply discarded the desk.

"Mr. K, Mr. K, Scott threatened to kill us!" I was brought back with a jolt from reading, "Into Thin Air," a book about an unsuccessful assault on Everest.

"What?" I stalled.

"He threatened to kill us," Miriam and Shannon repeated.

"Oh, I'm sure he was kidding. I'm sure he didn't mean it," I stalled further.

"Oh, no, he wrote it down."

Damn, I'd have to take some action. This being only a month or so after the Columbine Massacre, we'd been warned to take any and all threats with great gravity.

Scott watched my approach through heavily lidded eyes rimmed with red, not from crying but from barely repressed fury. Scott had been angry since the day he walked into my class, ejected by a school across town. He furtively closed his black binder. "What?" He pushed his tense body back in his seat, like a badger in a cage.

"You tell me."

"What've they been telling you?" He flared his nostrils at the girls, furrowed his brow, stared stonily at them in an attempt to look 'bad'.

"They said you threatened to kill them."

"If I was going to do it, I'd do it."

"Not here, you won't. Let's leave." A nervous silence hung over the class.

He didn't budge.

I reached over and opened his binder.

He grabbed it and clutched it to his chest.

I quickly wrested it from his grasp. It fell open to a page. Scrawled in black, clumsy cursive was a list of names under a heading, "Die!"

"Let's go. Now." I reached out to nudge his elbow up.

"I'm not going nowhere."

"Georgie, go get McGarry. Take this note." I pulled a notepad from my breast pocket and started to write.

"Okay. Okay. You win." He began to rise, pausing mid-glower to stare at me. "This time."

All of a sudden I stopped. I looked down at this frightened, angry little man. He looked awfully small, as if I were seeing him from a great distance. My indignation melted. What had we done to this little guy to create such hatred in him? How had our culture failed him and his family? How, in my participation in this petty passion play was I contributing to the delinquency of a minor, the further damaging of this little soul? And what, in the name of God, could I do about it?

"Scott," I whispered, my face softened from my confrontational mask, "Scott, let's go outside and talk. Please."

"All right," he grudged, "But I don't wanna."

"That's okay. Let's just go outside and talk."

I reached for him. He pulled away, but stood and trudged out the door, head lowered. I stopped at the threshold. "Silent reading. No talking. Don't get up. Don't." I scanned the class, letting the kids take in my solemn face. "Don't."

Scott's body language said everything he refused to say. Shoulders hunched, face pinched, seemingly taking great effort, eyes downcast.

I stood before him and looked around, surveying the beauty of the morning. The sky was still pink-tinged, the fog retreating over the ocean. Chestnut-backed chickadees skittered around the bird feeders we had hung in the garden, black hooded juncos scratched for spilled seed on the ground, ravens on the roof raucous in their scolding cries. I breathed deeply, exhaling long and slow through my nose as I scrabbled for a measure of calm.

"Scott," I said softly.

"What?" he responded angrily.

"Oh, Scott, you just can't do that."

"They made fun of me."

"Who? What'd they do?"

"They didn't do anything. They just gave me that look. I know what they were thinking."

"Scott, you can't do that. You can't frighten people that way."

His head jerked up, wild black curls highlighting his wild dark eyes. "You're not going to tell my mom, are you?" His frightened eyes widened.

Hmm. Another piece of the puzzle, I thought. "Scott, it's kind of out of my hands, man. You wrote it down, they saw it, they'll tell their parents. It's out of my hands. We have to do something about it."

"Please don't tell mom."

"Why not?"

"She'll be angry."

"What does she do when she's angry?"

He looked at me a long time. His tight mouth working soundlessly. "Just...don't," he pulled the words out of himself with great effort.

We started the long march to the office. Scott dragged his feet, head down. Dead man walking, I thought.

****

I suspended him. Over the vociferous objections of Scott's mom and the principal, I refused to readmit him until we all met with a counselor and came up with a game plan, a behavioral plan to deal with his deep resentment, distrust, and anger.

And we all made it through the weeks, the months, through the year in one fashion or another. Scott made a few friends, fleetingly. He always found fault with others, spread blame around liberally. Couldn't make true, lasting alliances.

On the last day of school, as I stood at the door giving final hugs, Scott hung back. Finally he came up to me and for the first time whispered a word of thanks, looked me in the eyes for a moment, bent down to pick up his backpack and walked out the door. I followed his slow, retreating form as he walked to his mom's waiting car.

As I watched, I saw him as he was, and as I feared he would be. I saw the figure of a very small child, lightly but ill-clothed, little fists clenched tight, carrying no tools and few skills, no one to follow, guideless and guileless, trudging off into the swirling mists, trudging ever so slowly through deepening drifts of snow, trudging off alone into the lengthening shadows. Not on a great adventure, not on an heroic quest, but simply because he couldn't stay, he just didn't belong.

Trudging off...into thin air.

Chapter 50
Those Who Can, Do.
Those Who Can't, Teach.

I didn't want to teach. My own experience in the Chicago Public School system did anything but whet my appetite for teaching. Teachers were paid so little they could not afford good clothes, most women dressed in sofa-style upholstery or in elderly floral prints, the few men were dull and drab, dressing like Soviet-style apparatchiks. Most rode public transportation to school because they could not afford cars. Even today, at a typical suburban high school, the parking lot with the older cars is the teachers' lot, the slick cars are in the kids' lot. Parental interest was rather anemic, running along the lines of, "If I even get a phone call from that school, you're going to get it!"

There were a few young women, but everyone knew they'd "find a husband" and get spirited away from schools, become housewives. Except, of course, for those poor old spinsters, those old maids, unable to find a mate and forced into teaching bondage for life, a life sentence, scratching and scrabbling to survive on what was called 'pin money'. 'Pin money' was extra money, a reference from a more frugal age, when, after staples were purchased, any money left could be spent on frivolity, such as sewing pins!

Sewing pins. Think about it. Teachers, especially elementary teachers, were not really expected to make a living teaching. If you didn't have a primary means of support, you were in trouble. Contrast that with the modern media culture. American kids grow up anticipating, demanding, an SUV, a home purchase in their twenties, an early retirement. "Wow, like life on the edge. Like no job to interfere with recreation (X-sports) and endless purchasing power. No fear, no hassles. Phat!"

Most school buildings were fortress-like, flat, featureless, forbidding (as they are today). Crest the hill approaching AnyTown, USA, and you are still greeted by edifices that may be factories, asylums, prisons or schools, take your pick. Some, like my high school, built the year after the great Chicago Fire of 1871, had ramparts and turrets for ease of defense against the rabble. Fronting on a four lane boulevard, Ashland Avenue, we had a complete absence of greenery, save a few courageous blades of grass battling their way through the pavement. Even in verdant rural counties, public schools use Round-Up and other herbicides to 'control' weeds (once again, control emerges as a dominant theme).

My wife and I, on a visit to Chicago 10 years ago, strolled by my elementary school as it was being readied for its 100th anniversary, its Grand Centennial. The cement had been power-washed, the bars on the windows freshly painted. The platoons of pigeons roosting in the gutters were the only occupants who looked at home. A typical American monument to education: solid, powerful, inflexible.

My thoughts trailed to California and the row upon row of featureless trailers hauled up from The Valley to 'warehouse' students, rafted out like junks or sampans in Singapore harbor. Consumables, devoured by mold, mildew and neglect. Disposables, like plastic diapers discarded and crumpled, strewn about the American landscape. As inhuman, as inhumane as the old fortresses were, at least they were well-built, with some character. These rows of trailers ('little boxes made of ticky-tacky' as Malvina Reynolds referred to them) are reflective of our age and our factory approach to schooling. The Bakersfield of Dreams.

And, reflective of the low status of teachers, always, always, the line: "Those who can, do. Those who can't, teach." Not even good enough to perform. Always a bridesmaid, never a bride, sleeping alone in cold, unconsummated beds. Always in the wings of life, never called on stage, "All right, Murray, this is your chance. Now teach!"

My 25th high school reunion was covered by a writer from the San Francisco Chronicle as part of a Sixties nostalgia thing. Ruthie (her real name) sat at the table of 'winners': doctors, lawyers, and the like. She ran down the list of successes then got to me, the lowly elementary teacher. "However, the Most Likely To Succeed, hasn't," she coyly observed.

So who, in his or her right mind, dreamed of becoming a teacher?

"Dad, when I grow up, I want to be a teacher and teach kids."

"Wise up, Conrad. Don't be a fool."

Chapter 51
Retirement

My wife and I showed up early for my meeting with the retirement counselor. Earline, the representative of the State Teachers' Retirement System, only rarely made the long, perilous trip out to the coast to meet with potential retirees, so I had scheduled my appointment several months before.

Cindy and I sat in the anteroom at the County Office of Education making small talk, nervous as hell about the prospect of major life change, anxious about the prospect of subsisting on a fraction of our income, with highly charged and mixed emotions about leaving a job, a calling, a mission I had come to love. My entire identity had become tightly entwined with my career. It was my calling card, my personna in the community, my cachet, my entrée into polite society.

But twenty years of personal and political struggle with an increasingly astringent hierarchy had left me battered and bloodied, if not unbowed. I continued challenging and caring for kids, keeping up a veneer of normalcy while the coils of the anacondic system tightened and tightened around my chest, choking my breath (taking the wind out of my sails, you might say) and threatening to constrict my heart.

It's well known that American education has been 'dumbed down' for years, suffering from mental atrophy, a turning away from higher level thinking. What is not as well documented, but a far greater ill, has been the loss of heart in this system. Programs have replaced interactions; systems replaced community. Curricula have been delivered, standards met (or not), while family and a culture of caring has been lost.

So I sit in the anteroom, awaiting my retirement conference, a rather disillusioned aging idealist, suffering from a more or less permanent state of head throb (etiology: persistent and cumulative effects of concussive interaction of head with brick wall), shortness of breath (etiology: constriction of stupid standards and stupider laws around chest), and heart pang (etiology: constant struggle with an uncaring bureaucracy).

"Mr. K, Mr. K."

I'm roused from my self-pitying little reverie by a young man strolling the corridor, stopping at our open door. Nattily clad in 3/4 length black coat, silk shirt, and black fedora, Nick, a student of mine from fifteen years ago, steps through the door to engage me.

"Nick, you look great, I mean, really good. What're you up to?"

Nick, a somewhat 'challenging' child from the distant past, looks remarkably clear, composed, and, well, grownup. He looks really together.

We catch up on old times, on new times, ranging from family to politics to the stupidity of the new High School Exit Exam (which he had passed as a freshman).

Suddenly he stops, cocks his head, and smiles broadly at me. "You know, Mr. K, the years have been good to you. You look just the same, well, a little whiter in the beard, but just the same. It's good to see you, man."

"It's great to see you too, Nick," meaning every word.

"What're you doing here?"

"I'm here to talk to a retirement counselor. It might be about time, Nick."

He responds with surprising fervor, "No, it oughta be against the law. No, not you, you can't ever retire."

I smile to accept the compliment. "Time and tide, baby. We'll see, we'll see."

We chitchat a bit more and part with a warm hug. Nick walks out the door, turns down the corridor, then backs up to the open doorway, again. Pointing his index finger at me as if it were an exclamation point, he says, "I just want you to know that you changed my life as a kid. I just want you to know that."

He held my eyes for a moment as I nodded acknowledgment, then he strolled off down the corridor again.

Cindy and I sat silently for a few moments, letting the event sink in. Gently she turned to me. "Why do you think that happened just now?"

As had happened over forty years earlier, in another lifetime, on a warm summer night as I sat on a limestone block in front of a temple in Chicago, watching a cockroach crawl over a carving, I realized yet again, "When the student is ready, the teacher will come."

CPSIA information can be obtained at www.ICGtesting.com
264203BV00001B/1/A

9 781432 706456